Endorsements for Ch

"As a lifelong advocate and scholar of ⌐ ⌐⌐⌐⌐⌐
as a woman from the global south, I have never before read a book
on the global climate crisis by a 'white American male' that has so
deeply touched and enriched me with its unprecedented combina-
tion of comprehensive scientific analysis, unswerving commitment
to justice and reconciliation, reverence for indigenous and margin-
alized people, honoring of women and the feminine."
—Dr. Rama Mani is Councillor, World Future Council and Convenor, Enacting
Global Transformation Initiative, University of Oxford.

"In his magnum opus, Choosing Earth, Duane Elgin leads us on
the ultimate journey of humanity's collective free will as a species.
Humanity, both present and future, and all life on Earth owes a
debt of gratitude to Duane for awakening us to the urgency and
regenerative possibilities of choosing Earth."
—John Fullerton is a former managing director of JP Morgan; founder of
Capital Institute, a key architect of "regenerative economics" that support
the transition to a just and sustainable future through the transformation
of finance.

"Duane Elgin taps into the most profound, evolutionary pulse of our
time. Choosing Earth is timely, relevant, clear, potent and abso-
lutely brilliant."
—Lynne Twist is the author of *The Soul of Money* and Co-founder of the
Pachamama Alliance, and a global visionary committed to alleviating poverty
and hunger, and supporting social justice and environmental sustainability.

"Having taken an oath to follow the Earth-honoring ways of
Peruvian shamanic wisdom, I consider what Duane Elgin has written
is of equal consciousness raising impact as that accomplished by
Rachel Carson when writing Silent Spring in 1962. Choosing Earth
may become vital reading among all dedicated to the Great Work
of safeguarding the sanctity and healthful flourishing of life upon
beloved Gaia-Pachamama."
—don Oscar Miro-Quesada is a traditional medicine carrier and healer from
Peru, Fellow in Ethnopsychology with the Organization of American States,
Invited Observer to the United Nations Forum on Indigenous Issues, author
of *Healing Light*.

"Duane Elgin's masterpiece is the most powerful and comprehen-
sive wake-up call on Earth. Choosing Earth invites us to take a hard
look at our current condition and possible futures, and sources

us towards a possible pathway of compassionate, conscious, and engaged Earth citizenship, rooted in elegant simplicity and fuelled by "aliveness". This passionate, eloquent and wise book is the critical curriculum humanity needs now to navigate its transition to a mature planetary civilization on Earth."

—Prof. Alexander Schieffer (Germany/France) is Co-Author of *Integral Development*, Co-Founder, Home for Humanity; and Co-Founder, TRANS4M Center for Integral Transformation.

"Choosing Earth *is a profoundly important book: the crowning achievement of a life of contribution by a man of enormous intelligence, extensive knowledge, penetrating insight, and far-reaching vision, who has dedicated his life to understanding the perils our civilization faces, and how we might survive them and even grow from them."*

—Roger Walsh, MD, PhD, teaches at the University of California Medical School and is a professor of psychiatry, philosophy, anthropology and religious studies. His books include Paths Beyond Ego and Essential Spirituality.

"*Duane Elgin's antennae are tuned to the deep future.* Choosing Earth *shifts our consciousness to the great awakening now required to save our planet and protect future generations."*

—Louie Schwartzberg is an internationally acclaimed cinematographer and director whose visual artistry celebrates life. His most recent film is "Fantastic Fungi," described as "a model for planetary survival" by the New York Times.

"*Duane Elgin's panoramic wisdom in* Choosing Earth *is vital in this time when complex, interconnected crises demand coherent, interconnected solutions. A pioneering and important book!"*

—Kurt Johnson, PhD, is an evolutionary biologist, interspiritual leader, professor of comparative religion at New York's Interfaith Seminary, and author of the award-winning *The Coming Interspiritual Age, Nabokov's Blues* and *Fine Lines*.

"*Duane Elgin is one of the foremost, visionary thought leaders of our time. His genius lies in his ability to combine emotional, intellectual and spiritual intelligences with his heart-based wisdom and knowing.* Choosing Earth *offers us a pragmatic guide to re-birth our planetary civilization that is otherwise destined to disappear."*

—Adam C. Hall is a successful entrepreneur, founder of the EarthKeeper Alliance, and CEO of The Genius Studio which empowers individuals and organizations to unleash their genius to discover and align their purpose.

CHOOSING EARTH

Humanity's Great Transition to
a Mature Planetary Civilization

*to Kelsey Anderson
With gratitude for your
work on behalf of the
well-being of all life.
Warmest regards,
Duane Elgin
9/28/20*

Duane Elgin

Published by Duane Elgin
www.DuaneElgin.com
Copyright © 2020 Duane Elgin

This book is part of the Choosing Earth Project.
For more information, go to www.ChoosingEarth.org

Book design and infographics: Birgit Wick, www.WickDesignStudio.com
Graph page 35: Emily Calvanese
Font: Georgia and Avenir
First Edition
1. SCI092000 2. NAT011000 3. OCC000000
ISBN 978-1-7348121-2-1

Contents

Preface

I have been developing much of the material in this book for decades and would like to introduce it with several reflections. Perhaps most important, exploring our pathway into the emerging future invites each of us into a deep self-initiation. We are being called to a uniquely personal engagement with our world in deep crisis unlike any in prior human history. Many will find this exploration shocking and alarming—as I certainly have. Therefore, *Choosing Earth* is intended for mature individuals who have the inner resilience to look into our world as it moves through a profound rite of passage and to discover extraordinary opportunities, now largely hidden within the tumult of our times.

This book is written for embodied warriors of the heart—persons who have the inner wisdom to explore new ways of relating to the world and the outer zest for life to bring their gifts into our time of great transition. A new culture and consciousness are emerging, calling each of us to a new level of maturity and engagement. *Choosing Earth* dives deep to explore inner changes in who we are and outer changes in how we show up in the world. Our time of great turning is not a problem to be fixed—instead, we have entered a rite of passage into a new understanding of reality, human identity, and evolutionary journey.

If these themes speak to you, I hope you will join me on this demanding and enlivening journey of choosing Earth—she is always welcoming us home with feelings of belonging and wonderment.

Duane Elgin, March 2020, Woodacre, California

PART I:

∙∙

Introduction to Transition

"I used to think the top environmental problems were biodiversity loss, ecosystem collapse and climate change. But I was wrong. The top environmental problems are selfishness, greed, and apathy . . . and to deal with those we need a spiritual and cultural transformation—and we scientists don't know how to do that."[1]

—Gus Speth, former director, Council on Environmental Quality

". . . we are in a state of planetary emergency: both the risk and urgency of the situation are acute. . . the intervention time left to prevent tipping could already have shrunk towards zero, whereas the reaction time to achieve net zero emissions is 30 years at best. Hence, we might already have lost control of whether tipping happens. . ."[2]

—Tomothy Lenton, et. al., "Climate tipping point"

After tens of thousands of years, humanity has begun to move into a time of "great transition." Early in the decade of the 2020s, humanity's long journey of development is awakening to the reality that we have reached a time of historic transition, unprecedented in its urgency, scale, and severity. All at the same time, the human community confronts a multitude of accelerating crises; for example, increasing climate disruption, spreading regions of water scarcity, declining agricultural productivity, growing inequality in wealth and well-being, rising numbers of climate refugees, spreading extinction of plant and animal species, drowning in oceans polluted with plastics, and aging bureaucracies of overwhelming scale and complexity.

Taken together, it is clear we have reached the limits of the Earth's capacity to sustain our current trajectory of growth. I do not regard this predicament as an evolutionary mistake. Instead, we are behaving in a predictable manner, no different from the rest of life.

As personal background to this book, since 1978, I have been writing and speaking about the decade of the 2020s as the time when humanity would hit an "evolutionary wall" and begin making a pivotal turn as a species.[3] To clarify, an *ecological* wall" emerges when we run into the physical limits of nature to sustain humanity. In contrast, an *evolutionary* wall" emerges when we run into ourselves—when we run into the limits of our adolescent behavior and are pushed to turn toward more mature ways of being and living on the planet. An evolutionary wall presents humanity with an identity crisis as great as our ecological crisis: Who are we as a species and what is the larger journey we are on?

Only after years of inquiry did I conclude that, if humanity is to have a promising future, we must choose a different pathway for living on Earth. Our current trajectory is driving us into an unyielding ecological wall and disastrous future. We are challenged to wake up together and rise to a higher maturity and responsibility as a species.

This shift will not happen automatically: We either consciously choose, together, a regenerative path for living on the Earth, or we lose our future. We either choose the Earth as a welcoming home or else continue our rush toward ruin.

Choose it or lose it.

This stark understanding emerged for me after years of futures research—including: two years of work as a senior social scientist with the Presidential Commission on the American Future, looking ahead from 1970–2000,[4] plus six years of work with the "futures group" at the Stanford Research Institute (now SRI International). There I co-authored the study "Changing Images of Man" with Joseph Campbell and a small team of scholars, looking at the deep archetypes drawing us into a more promising future.[5]

Other major futures studies included a long-range report for the National Science Foundation and President's Science Advisor,[6] exploring the inertia of large, complex bureaucracies and their inability to respond rapidly to disruptive change.[7] Another study explored alternative futures for the Environmental Protection Agency for the period 1975–2000.[8]

In response to what I was learning, I began writing about and struggling to practice a life of "voluntary simplicity."[9] After leaving SRI in 1977, I invested more than a decade in community organizing in the San Francisco Bay Area and beyond to create large scale, "Electronic Town Meetings" using broadcast television combined with telephone-based voting from a scientific sample of citizens. Our objective was to demonstrate that people at a metropolitan scale can have a powerful voice in choosing our future.[10]

Overall, these many learning experiences led me to the conclusion that the decade of the 2020's would be pivotal for humanity's future—a time when we would be confronted with unyielding questions of who we as a species and what kind of journey we are on.

Now, more than 40 years later, the fateful decade of the 2020s has arrived. For me, the preceding years have been profoundly challenging—physically, emotionally, mentally and spiritually. Absorbing the scope, speed and depth of change required to respond realistically to the world unfolding has been deeply confronting. Sorrow has been my faithful companion—anguish my teacher. I've been humbled by the intensity and immensity of suffering growing in the world, knowing this tsunami of sorrow will break our hearts and, at the same time, open us to our higher humanity. Although writing has been a major part of my occupation, this journey has been a soulful challenge beyond words. My writer's desk has become an altar to despair as I accept all that will perish.

As I step back to look, again and again, seeking perspective for what is unfolding, I recognize that I am writing this book from a privileged perspective—that of a white, male member of a highly industrialized, Western culture and nation. Although my roots are in a small farming community in Idaho, I have lived most of my adult life in a modern, urban-industrial culture. Yet, as I try to find my place in a world in profound transition, I find myself returning to my experience as a farmer.

Now I see myself planting seeds of possibility but without expectation that I will live to see them blossom in a new summertime or partake of their fruits in the harvest of a distant autumn. My approach now is to trust the wisdom of the Earth and the human family in bringing forth another season of life.

We are creating a rite of passage for ourselves as a species—but a passage to what? Could the enormity of imagined loss be the catalyst for unimagined gain? Could a new human alloy, rich with aliveness, emerge from the furnace of the superheated decades we have now entered?

In stripping away the trivial in search of the essential, I realize what matters most is life itself.

Aliveness is our only true wealth.

1. Our World in Great Transition

Reaching limits to growth is a recurring and predictable development throughout nature. Most species commonly seek to exploit their ecological niche to the fullest. Humanity's "niche" or operating domain is now the entire Earth—and we are exploiting it to the extreme.

Overshoot and collapse are common in nature. We humans are not a special species, immune from the lessons experienced by the rest of life. We learn through experience and we have never before encountered this situation. So, we should not be surprised if a great tragedy is necessary to awaken the evolutionary intelligence of humanity. Great compassion will be needed to cope with the immense suffering that will result from our technological powers combined with our limited experience.

The suffering unleashed by our times of great transition is having a disproportionate impact on poor people, indigenous people, and people of color. If we respond to these transitional times with our collective understanding, compassion, and creativity, we have the potential to liberate ourselves from our limited sense of humanity and awaken to a new sense of species identity.

Despite all our good intentions, without this coming era of unprecedented suffering and adversity, the human family seems unlikely to awaken to our species maturity and evolutionary opportunity. The immense suffering of billions of precious human beings may burn through our complacency and sense of separation. Suffering is the fire that can awaken our compassion and fuse humanity into a cohesive, global civilization. We are discovering the circle of life is closed and there is no escape from the limits to unrestrained growth. Like it or not, we must now recognize the Earth is a single, tightly interconnected system that is measurably breaking down and moving toward collapse. We should not be surprised we have reached a predictable overshoot of the Earth's ecology. What is not predictable is where we go from here.

Looking at current trends, it seems reasonable to conclude that dark times are ahead. Have we reached the end of our evolutionary journey—a brutal and degrading conclusion to our awakening and growth as the dominant species on the Earth? Have we already wrecked our chance of realizing our higher potentials as a species? Or, are there other possibilities not yet widely recognized? Could humanity be going through a profound rite of passage and making the transition to a higher maturity with a new understanding of ourselves and our journey? This points to a key possibility missing from many contemporary conversations about our future:

Unyielding challenges
contain unprecedented opportunities.

An emerging whole-systems crisis provides an opportunity to enter a new pathway of awakening, development, and discovery. A number of people have written about our time of great transition.[11]

For example, eco-philosopher, spiritual activist, scholar, and elder, Joanna Macy, has described this as the "great turning"—a transition from an era of urban-industrial growth to a life-sustaining world. How we navigate this pivotal rite of passage will be key to the world that emerges.

At the beginning of the decade of the 2020s, we are between worlds. The old world is unraveling and, at the same time, a new world is being woven together. We face big challenges and it will take an equally big vision to transform conflict into cooperation and draw us into a transforming future. The push of necessity is so great—genuinely bringing planetary civilization to a place of collective breakdown and disintegration—that we can easily overlook a corresponding pull of opportunity.

The most difficult challenge facing humanity is not devising solutions to the energy crisis or climate crisis or population crisis. Rather, it is bringing narratives and stories of the human journey into our collective imagination that can empower us to look beyond a future of great adversity and see a future of great opportunity.

2. Stories of Transition

Without a story or vision of a promising future, we are lost and can create a ruinous future for ourselves. For example, the "American Dream" that pulled the US, and a good portion of the world, forward for generations is fast becoming the world's nightmare as the excesses of consumerism produce climate disruption, species extinction, and enormous disparities of wealth.

Our story is our future!

Now, instead of a different "dream," people need wide-awake visions of a purposeful and sustainable future told in ways that are believable and compelling. To be effective, we require a compelling narrative that speaks to the entire human family and to do that it must be:

Simple—able to be told in just a few words

Universal—understood by everyone on the planet,

Emotionally powerful—so people feel and care about it, and

Evocative of our higher human potentials—a story that calls forth our gifts and explains how we are on a purposeful journey as a species.

We are just beginning to discover this story and to recognize core elements in our personal lives. In this rare moment in human history we are beginning to develop, for the very first time, the "story of, by, and for all of us." To illustrate, here are four widely recognized narratives we can use to portray the human adventure:

1. Humanity Is Growing Up. Over tens of thousands of years, the human species has been learning and maturing. We have moved from our childhood as awakening hunter-gatherers to our late adolescence as a species that is on the edge of a planetary civilization. We are now moving through a collective rite of passage, toward our early adulthood as a human community.

In shifting from our species adolescence to early maturity, we are building a new relationship with the Earth, one another, and the universe. This story is such an important one, offering deep insights for our time of great transition, that I will return to explore at length later in this book.

2. The Global Brain is Waking Up. Our ability to communicate enabled humans to slowly progress from nomadic bands of gatherers and hunters to the edge of a planetary civilization. Now, within the space of a single lifetime, the human family has moved from the separation of geography and culture to nearly instantaneous global communication and connection. With stunning speed, we have developed tools of local to global communication that are transforming our collective communication *and consciousness* as a species.

The "global brain" is a metaphor for the worldwide network formed by people coming together with communication technologies that connect us into an organic whole. As the internet becomes faster, more intelligent, more ubiquitous, and more encompassing, it increasingly ties us together in a single communications web that functions like a "brain" for planet Earth.

Because we are in the midst of an unprecedented revolution in the scope, depth, and richness of global communications, the impact of this revolution on our shared consciousness is equally unprecedented. No longer isolated and cut off from one another, we are collectively witnessing our world in profound transition. Awakenings and innovations happening on one side of the planet are being communicated instantly around the world, enabling us to wake up together. With astonishing speed, we are rousing from our slumber to know ourselves as a single species, united by an extraordinary network of planetary communication.

3. The Planet is Giving Birth. Storytellers from around the world have long used the birthing metaphor to give meaning to suffering and transformation: The pain is worth enduring for the sake of what is being born. A powerful description of birth is taken from Dr. Betsy MacGregor's book, *In Awe of Being Human: Stories from the Edge of Life and Death.*[12]

"[B]ringing a new human being into this world is not an easy matter. The ordeal that must be endured is huge, and it can take a significant toll. The life of either mother or child, and sometimes both, may be lost if all does not go well. . . . Tremendous forces must be set in motion in order to expel the infant from the comfort of the womb, and the going can get extremely rough. Powerful maternal muscles create rhythmic waves of contraction that force the baby along, while the youngster's head leads the way, stretching apart tight maternal tissues and pushing past rock-hard bone, bearing the brunt of the work. The amount of pressure exerted on the infant's head is so great that the soft bones of its skull are squeezed hard against each other and made to overlap, only slowly regaining their normal position days after birth. Collections of blood may form in the infant's scalp from the battering as well. For hours upon hours the formidable process goes on, testing the limits of endurance for both mother and child. It's enough to make an observer exclaim, 'Good Lord! Why has nature made it so hard for us to enter this world?'"

We may cry out with similar exclamations regarding the birth of our species-civilization. "Why is it so hard for us to wake up together and enter this world as a whole species?" A precious miracle is hidden within the pain of our time of great transition—a whole new life is being born. When we know a new life is emerging on the other side of the pain and unstoppable push for the birth, we are able to move ahead with less fear and with loving anticipation. Jacques Verduin offers deep insight into how we can bear the suffering of this new world being born:

"*Amor* means love. *Fati* means fate. *Amor fati* is a Latin expression that means loving your fate, including suffering and loss. The key to achieving *Amor Fati* is an attitude of deep acceptance of the events that take place in your life. It is the practice of embracing what happens, particularly the painful things. You don't have to like what is going on and you don't have to agree with what's going on for you to be able to accept it. The fact that a situation is presenting itself simply means you get to have the opportunity to show up for it and accept that this situation is happening. . . There is an art to learning how to bear one's suffering. The word origins of the verb 'to bear' relate to the verbs 'to endure' and 'to give birth.' Perhaps bearing our suffering refers to enduring and sustaining pain in such a way that, ultimately, through our labor, it gives birth to a new and vital realization, a new state of consciousness that offers strong guidance on how to live."[13]

4. The Sacred Feminine is Re-emerging. From at least 50,000 years ago until roughly 6,000 years ago, an "Earth Goddess" perspective provided the primary understanding for the relationship of humans with the larger world.[14] The feminine archetype recognized and honored the aliveness and regenerative powers of nature and the fertility of life. Then, roughly 6,000 years ago, with the rise of city-states, more differentiated classes (priests, warriors, merchants), and more complex cultures, a masculine mindset and a "Sky God" spirituality became dominant and supported the development of human society organized into larger-scale structures and institutions.

A masculine, patriarchal mindset has grown and developed over thousands of years and has encouraged the growing individuation, differentiation, and empowerment of people. It has also supported humanity's growing separation from and exploitation of nature that has led up to our current time of crisis and need for transition. Perhaps the last vestiges of this mindset are being expressed in hyper-masculine leaders who are seeking to retreat into isolationism, nationalism, and a renewed emphasis on materialism and authoritarianism. Nonetheless, a new mindset is now emerging with the reawakening of a feminine perspective that regards the Earth—and the universe—as a unified, supportive, and regenerative organism.[15]

Bringing these four narratives together, we can already begin to tell a compelling story that gives new meaning to our time of transition:

> *Humanity is growing up and moving toward*
> *early adulthood, the global brain and*
> *human consciousness are waking up,*
> *and a species-civilization is being born that*
> *embodies the relational and nurturing*
> *perspective of the deep feminine.*

With these narratives, we can begin to visualize a meaningful journey ahead. In turn, if we can imagine a common journey into a purposeful future, we can build it.

With a common story, we can see our supporting roles, our lives become more meaningful, and change less overwhelming. In this vulnerable time, it is vital for the stories we tell to be worthy of our time of great transition. If we are without a compelling story and feel lost, then it is easy to be frightened and to focus on threats to our security and survival. New energy policies, for example, do not begin to go deep enough to change our species mind and overall sense of direction. However, a powerful story of the human journey can provide the social glue to pull us together in common effort and take us in a regenerative direction.

Although the breakdown and unraveling of the current world system seems inevitable, it need not represent the end of our journey but rather a stage of fierce transition. From this larger perspective, humanity seems midway along an infinitely larger journey than we previously imagined. At the same time, there is ample reason to think we are headed for a climate catastrophe within this century that, without a larger story to guide us, could produce a largely uninhabitable Earth.

These are challenging times for all of humanity, whether impoverished and oppressed or wealthy and privileged. Although the specific challenges we each face may vary greatly, the overarching challenge of this planetary moment is shared among us all.

The challenge of moving beyond a time of great transition and establishing ourselves as a viable species-civilization is the primary focus of this book.

It is abundantly clear that the more privileged people of the Earth must rapidly transform nearly every aspect of their lives if some form of viable species-civilization is to emerge: the energy we use, the levels and patterns of consumption we choose, the work we do and skills we develop, the homes and communities in which we live, the food we eat, the transportation we use, the education we acquire, and the way we treat people who are of different races, genders, cultural and sexual orientations, generations, and more.

The wealthy and powerful may shield themselves temporarily, but not permanently. Eventually the privileged minority will inevitably be forced to address the fact that we cannot continue our current ways of living and still hope to have a habitable planet.

If we do not step up to meet both the material and social challenges of our times, then we are sure to follow the example of more than twenty great civilizations that have collapsed through history; including Roman, Egyptian, Vedic, Tibetan, Minoan, Classical Greek, Olmec, Mayan, Aztec, and a number of others. Our vulnerability is made starkly evident as we recognize the breakdown and disintegration of these great civilizations of the past. However, the current situation is unique in one key respect—human civilization has reached a global scale and encircles the Earth as an interdependent system. The circle has closed. Now the simultaneous downfall of *all* the intertwined civilizations on Earth is threatened. How can we even contemplate this devastating prospect?

An extraordinary push and unprecedented pull are at work in these transitional times. If we look only at the push and ignore the pull, it places our journey in great peril. To visualize this process, imagine pushing on a length of string. Pushing ahead, the string will bunch up in front of us and create a tangle of knots. Then imagine simultaneously pulling on the string—it no longer bunches up in a jumble but can move forward in a line of progression. In the same way, if we understand and respect both the pushes and pulls of our times, we can move ahead without getting completely entangled in the process.

If we take into account only the unyielding push of the climate crisis combined with other adversity trends, then our efforts will produce complex knots and we can easily become mired in confusion and despair. However, if we deepen our vision to include the pull already present, we have the potential to move ahead swiftly on a remarkable journey as a species.

The pull of opportunity does not eliminate the enormous challenges we face. Instead, it balances our approach by helping us see beyond immediate challenges to a larger vision of possibility. By recognizing and working with both the powerful push of necessity and the remarkable pull of opportunity we can find the courage, compassion, and creativity to work through the difficulties of transition.

We are challenged to look beyond the ecological devastation of the Earth if we are to rise to a new maturity and consciousness essential for building an enduring planetary civilization. If we are successful in this transition, neither the Earth nor humanity will be the same. To explore this great transition, it is helpful to take a whole systems view with three perspectives:

Look Wide: Look beyond single factors such as global warming and include a wide range of trends and consider them as a whole system: climate disruption increasing, population growing, climate refugees migrating, resources depleting, species dying, inequities increasing, artificial intelligence emerging, and much more. Looking wide provides us with a clearer picture of change and brings an integrative perspective that is missing from a singular focus.

Look Deep: Look beneath the outer trends such as climate change and species extinction to include the inner realities of evolving psychology, values, culture, consciousness, and paradigms. Include a regard for the invisible realms of psyche, soul, purpose and ethics.

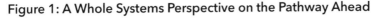

Figure 1: A Whole Systems Perspective on the Pathway Ahead

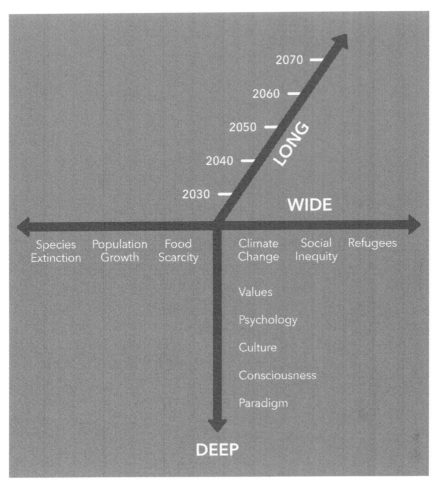

Look Long: Look far into the future—much farther than the short run of the next five or ten years. Trends that are uncertain and ambiguous in the short run become much clearer when extrapolated to the longer run where their impact on future generations becomes more apparent. Therefore, this inquiry looks a half-century into the future where trends become much more distinct and well-defined.

When we look wide, deep and long, we begin to see more clearly the pivotal time in history we have entered and how we can move more deliberately beyond our time of great transition. Before exploring the stunning potentials that draw us forward, it is essential to acknowledge the extraordinarily powerful pushes that are driving us to discover a new pathway ahead.

3. Warnings to Humanity

Many people know we confront Earth-scale challenges. We can already feel the leading edge of a gathering world storm—a whole-systems crisis enveloping the entire Earth. Although human societies have confronted major hurdles throughout history, the challenges of our era are unique in one crucial respect: most are planetary in scope and there is no escape. Like it or not, the Earth has become a single, tightly interconnected system. Never before has humanity confronted a crisis that is devastating the entire biosphere and crippling the ecological foundations for all life.

A summary evaluation of humanity's crisis of transition was developed by the world's leading scientists decades ago. This assessment was offered in 1992 when over 1,600 of the world's senior scientists, including a majority of the living Nobel laureates in the sciences, signed an unprecedented document titled *Warning to Humanity*.[16] In their historic statement, they declared that "human beings and the natural world are on a collision course . . . that may so alter the living world that it will be unable to sustain life in the manner that we know." This is their warning:

> "We, the undersigned senior members of the world's scientific community, hereby warn all humanity of what lies ahead. *A great change in our stewardship of the earth and the life on it is required if vast human misery is to be avoided and our global home on this planet is not to be irretrievably mutilated.*"[17] [emphasis added]

This conclusion is in accord with my views based on a half-century of research and writing.[18] My thoughts return to a key portion of this warning. Scientists state that, if great changes are not made in our stewardship of the Earth, the planet will be "irretrievably mutilated."

It is the last, two words
—"irretrievably mutilated"—
that reverberate in my being.

What do these two words mean for countless generations ahead? The Earth forever disfigured, permanently damaged, maimed and mutilated for all time? Is that to be our legacy of failed stewardship to future generations?

A 25-year update to this famous warning was published in 2017 and concluded that the initial warning to humanity had not been heeded—and then added this further warning:

> *"Soon it will be too late to shift course away from our failing trajectory, and time is running out.* We must recognize, in our day-to-day lives and in our governing institutions, that Earth with all its life is our only home."[19] [emphasis added]

This is a stark update to the warning from the world's senior scientists. A further warning from thousands of scientists was issued in 2019, stating:

> "Scientists have a moral obligation to clearly warn humanity of any catastrophic threat and to 'tell it like it is.' On the basis of this obligation and . . . [scientific evidence] we declare, with more than 11,000 scientist signatories from around the world, *clearly and unequivocally that planet Earth is facing a climate emergency."*[20] [emphasis added]

There are compelling reasons for these blunt conclusions. Global warming is now our most visible challenge and is speeding ahead. We are on the verge of crossing a climate threshold beyond which global warming will feed upon itself and produce runaway changes. If global warming gets out of control, we will have lost control of our future. The time-frame for reaching this threshold is shockingly close—perhaps a decade or two.

These warnings are not new—we have been hearing about threats to our future for a half a century or more. For example, the eminent biologist Rachel Carson, said this in speaking to college students in 1962:

> "The stream of time moves forward and mankind moves with it. Your generation must come to terms with the environment.

You must face realities instead of taking refuge in ignorance and evasion of truth. Yours is a grave and sobering responsibility, but it is also a shining opportunity. You go out into a world where mankind is challenged, as it has never been challenged before, to prove its maturity and its mastery—not of nature, but of itself. Therein lies our hope and our destiny."[21]

Because these warning give much attention to the themes of "breakdown" and "collapse," I want to clarify them at the outset. These terms are often used interchangeably but can be understood quite differently:

Breakdown means that linkages in key systems are failing. Power outages occur. The water stops running at times, and its purity is doubtful. Fire and police departments close periodically because they cannot pay people. Supply chains of food delivery—from farm to stores—stop operating for periods of time. Breakdown means the disintegration of whole systems into their component parts which, while deeply disrupting and damaging to health, employment, and access to necessary resources for many individuals, also creates opportunities for new configurations of living. By disrupting business as usual, breakdowns create openings for rebuilding in new ways that can be healthier and more resilient. Breakdowns can be a catalyst for people's creativity and spur innovation—for example, in building and retrofitting communities whose local economies support resilient approaches to living.

Collapse is far more serious than breakdown as it describes the ruinous downfall of communities, cities, and civilizations. With collapse, society fails completely as housing, transportation systems, water and sewage systems, and more fall into jumbled chaos. Collapse is the catastrophic failure of the system *and* its components. Collapse leaves both in a condition of rubble—a junkyard of broken systems of all kinds—transportation, communication, and civic services. Collapse produces a very difficult foundation (physical, economic, psychological, social, and spiritual) from which to build a promising future of inclusive, sustainable well-being.

A graphic description of what collapse could mean for the world is offered by what is happening in Venezuela. Once one of the economic miracles of South America with one of the largest reserves of oil in the world, its economy has collapsed in the last few years with devastating consequences:

" . . . Desperate oil workers and criminals are stripping the oil company of vital equipment, vehicles, pumps and copper wiring, carrying off whatever they can to make money. .. Venezuela is on its knees economically, buckled by hyperinflation and a history of mismanagement. Widespread hunger, political strife, devastating shortages of medicine and an exodus of well over a million people in recent years have turned this country, once the economic envy of many of its neighbors, into a crisis that is spilling over international borders."[22]

With *breakdowns*, the components of life are still sufficiently intact to be re-assembled into new configurations that can work—hopefully even better than before. However, *collapse* requires building a new world on the scrap heap of ruined infrastructure, shattered institutions and a devastated biosphere.

In addition to the catalogue of dangers already described, other risks loom large; for example, the possibility of nuclear conflicts producing a devastating nuclear winter; systems of artificial intelligence that escape human control; genetic engineering producing an array of human species not friendly to "ordinary" humans; a major asteroid strike, and more.[23] Although these threats are of obvious importance, they are not emphasized here. Instead, the focus is on extreme climate disruption and a global systems crisis producing a planetary scale breakdown and collapse.

4. The Magnitude, Speed, and Depth of Change

The astonishing magnitude of change required to respond in ways commensurate with the challenges we face is nearly inconceivable. The editors of the respected journal, *New Scientist*, offered this assessment of the work that lies ahead for humanity:

"It will arguably be the largest project that humanity has ever undertaken—comparable to the two world wars, the Apollo program [to put a human on the moon], the cold war [with a nuclear arms race], the abolition of slavery [which included a civil war], the Manhattan project, the building of the railways and the rollout of sanitation and electrification, all in one. In other words, it will require us to strain every muscle of human ingenuity in the hope of a better future, if not for ourselves then at least for our descendants."[24]

To step up to this level of change is completely unprecedented and will require nothing short of a revolution in collective effort by humanity. However, even this stunning description fails to reveal the depth of practical change that is essential. In particular, a sweeping transformation of energy production and use is immediately required. To avoid disastrous global warming, scientists estimate that the human community must halt increases in fossil fuel emissions in 2020, and then cut them in half by 2030, and then cut them in half again by 2040, and then come to a net zero of carbon emissions by 2050.[25] The *entire world* must either eliminate or offset carbon pollution by mid-century. This means that:

- By 2050, no home, business, or industry will be heated by gas or oil or, if they are, their carbon pollution must be offset;

- No vehicles can be powered by diesel or gasoline.

- All coal and gas power plants must be shuttered.

- Even if the world succeeds in generating all its electricity from zero-emissions sources, such as renewable energy or nuclear power, electricity makes up less than one-third of current fossil fuel consumption. Therefore, other energy-intensive users of fossil fuels—particularly those used to manufacture steel and concrete—must be fueled by renewable sources.

While a complete rebuilding of the entire energy infrastructure of the world within a few decades is vital for a workable future, it is far from sufficient. In addition, a deep and profound transformation is required in virtually every aspect of life—the food we eat, the skills we develop and the work we do, the homes and communities in which we live, the media messages we produce and receive, the local to global conversations we develop, the values of economic fairness and social justice we share, the leadership offered across diverse institutions (political, religious, media, non-profit), and more.

An entirely reconfigured society, economy, culture,
and consciousness is our only pathway
for avoiding the irretrievable mutilation of the Earth.

A stark appraisal of the depth and breadth of economic changes is described in the study, "Life Beyond Growth":

> "Can the world's nations, corporations, cities, institutions, and households convert their economies to green ones, where the net impact on nature is within the boundaries of what the planet can sustain? As challenges go, this one looks enormous: . . . a country like Japan would have to cut its consumption of resources and environmental impact by (very roughly speaking) more than 50%, while the United States would need to reduce by a factor of 75%."[26]

Responding to the climate crisis requires a dramatic reduction in the levels and patterns of consumption. The recently acquired habit of overconsumption is pumping huge amounts of CO_2 into the atmosphere in order to meet current production and consumption demands, particularly those of the wealthiest 20% of humanity. When we calculate the human footprint on the planet—a measure of how intensively we are using the Earth's natural resources—we find that:

> "It takes a year and a half to generate the resources that the human population uses in only a year. . . Another way to look at this is to say that it would take 1.6 Earths to produce all the renewable resources we use [in one year]. And worse, the human population is expected to use the equivalent of 2 Earths of renewable resources per year by 2050. The effect of this overuse is a growing scarcity of resources. . ."[27]

We have already exceeded the Earth's limits to growth.[28]

Undeniably, we humans have a completely unsustainable way of living that is based on people in the wealthier countries and regions consuming far more than their fair share of the resources.

This overconsumption is depriving a majority of humans from their fair share and condemning them to poverty and a disproportionate level of climate-induced suffering. This inequity is so extremely discriminatory and unbalanced that it cannot endure.

How can we implement a massive and complex transformation in living to bring us into balance with nature's limits? It will be immensely challenging for those with high consumption lifestyles to deliberately limit their drawdown of resources and to share their wealth with those less economically privileged. Humanity's survival requires a lifestyle revolution where the wealthy choose ways of life that are materially far more restrained in the use of the Earth's limited resources and far more generous in fostering the well-being of those who are poor.

This shift is more than a matter of moral justice and fairness. It is also essential for preventing all-out class warfare over resources and for reframing our relationship with each other and the rest of life. If we are going to work together as a human community, then those who are accustomed to being in positions of authority and power (as a result of class, gender, race, geography, age, ability, education, etc.) must step forward to lift up the voices of the global majority (poor folks, indigenous communities and other long-suffering and oppressed groups). Only then will it be possible to create meaningful, systems-level changes, including the redistribution of resources that will liberate the global majority from being forced by survival pressures to focus only on their urgent, short-term needs.

In addition to great concern for the magnitude of change, alarm is growing with regard to the speed of change, particularly with respect to climate disruption. In the past, scientists thought it would take centuries if not thousands of years for climate to swing into a different configuration. It came as a profound shock for scientists to discover *a profound shift can occur within "a matter of decades or even less."*[29] To illustrate, a period of global cooling called the Younger Dryas occurred roughly 11,800 years ago (likely the result of an asteroid breaking up in the atmosphere) and was followed by a period of abrupt warming, estimated to be roughly 10°C within a matter of years![30]

Although such astonishingly rapid levels of temperature change are not currently predicted, this example does reveal our vulnerability if we ignore historical variations. Current institutions of government and policy thinking would be completely unable to step up to such abrupt climate change.

Most governing institutions are designed to perpetuate the past, not to move swiftly into a transforming future.[31]

To make the rapid changes that are essential will take unprecedented levels of personal and institutional responsiveness and responsibility. Can we do this?

A personal story illustrates the unresponsiveness of government bureaucracies to recognized threats to our future. I first learned about global warming as an existential threat to humanity in 1975 while working as a senior social scientist on a year-long project for the President's Science Advisor at the think tank, SRI International.[32] I was part of a small team looking for unexpected, future challenges that could wipe us out from the blind side.

In support of this project, I attended a briefing on climate change at the Department of Energy in Washington, D.C. At the briefing we were told that, if present trends continued with CO_2 build up, in another 40 to 50 years it would create serious problems of global warming for the planet. Despite this somber warning, energy officials discouraged us from including global warming in our report. They reasoned that this would not grow into a crisis for nearly half a century and this would give the political process plenty of lead time to mount a response. Not only did we not include global warming in our report, the sponsoring government officials decided the report was too controversial for public dissemination and it was put on the shelf, away from easy access by politicians and the public.

Now, nearly a half-century later, we can see the result of decades of delay: As predicted, the world is under assault with a dramatically changing climate and the unraveling of civilizations. Given this kind of experience, I do not expect existing institutions—in government, business, media, and education—to rise quickly to the unprecedented challenges we now face. As I wrote in the report to the President's Science Advisor: our large, highly complex bureaucracies are not configured to respond with the speed and creativity necessary to meet the challenges of our perilous times.[33] For that reason:

I place my greatest faith
in the people of the Earth organizing ourselves
from the local to global level and, together,
swiftly learning our way into
a sustainable and purposeful future.

In addition to the magnitude and speed of change, we also are obliged to recognize the depth of change required to meet our times of great transition. Recall the kind of change needed and described powerfully by Gus Speth, former chairman of the Council on Environmental Quality and co-founder of the Natural Resources Defense Council. He said:

> "I used to think the top environmental problems were biodiversity loss, ecosystem collapse and climate change. But I was wrong. The top environmental problems are selfishness, greed, and apathy . . . and to deal with those we need a spiritual and cultural transformation—and we scientists don't know how to do that."[34]

No one alive has ever gone through the fiery rite of planetary passage into which we have now entered. Engaging in this great transition takes emotional courage as it naturally brings up existential fears and deep-seated wounds. The sheer magnitude, complexity, and turmoil created by world-changing trends produces feelings of discouragement and despair that can be a barrier to looking ahead. However, denial of crisis is also a denial of opportunity.

We can no longer turn away. Despite being pushed by Earth-changing forces, we are challenged to look ahead with heartfelt care and maturity. The alternative is to surrender the future and descend into chaos and ruin.

PART II:

Three Pathways Ahead

"We are pilgrims together, wending through
unknown country, home."[35]

—Father Giovanni (1513)

5. Collapse, Authoritarianism, Great Transition

Before focusing on a pathway of great transition to a more promising future, it is important to acknowledge how open and vulnerable the future is for humanity at this singular moment. We have entered an extraordinarily rare interval in human history—a choice point in humanity's collective journey—a space between the past and the future where the path for generations ahead, depends on choices we make now. We cannot predict where humanity will go from here for a simple reason: Our future depends on our conscious choices—or failure to choose—both individually and collectively. Our evolutionary journey will either become conscious of itself or descend into darkness. This is a pivot point in human history—a time that will forever be remembered as we either rise in our maturity as a viable species-civilization, aware of our responsibilities and profound interdependence, or fall back into ruin.

These stark choices did not have to be our destiny. Humankind squandered its opportunity for a pathway of gradual adaptation nearly a half-century ago, in the 1970s, when the immense challenges we face today were first recognized. It is now too late to choose a path of gradual change. We have consumed the margin of extra time—at great cost—to keep the status quo alive for a few additional decades.[36]

Having used up the breathing room for gradual adaptation, humanity confronts catastrophic consequences if we don't respond swiftly to make sweeping changes in how we live on Earth. Within a few decades, large portions of the planet will no longer be fit for human habitation. Extremes of drought, floods and storms will become common. Famine and disease will shake humanity to the core. Hundreds of millions of climate refugees will be on the move, looking for places to live. The mass extinction of animals and plants will forever impoverish the ecology of the Earth. Recognizing all this, we realize the options now open to us are starkly limited.

The window of opportunity for gradualism has closed.
We either step forward decisively
or descend into chaos or authoritarianism.

Below are three major pathways ahead that are present in the world (see Figure 2 on page 35). Each of these can emerge from the same underlying trends. The difference is not in the driving trends but in the choices we make in response to those trends. *Although all three pathways are likely to continue to varying degrees throughout the next half century, the pivotal question is which of the three will become dominant in guiding us into the far future.* Described simply, these three pathways are:

Pathway 1: Chaos, crash and collapse. In this pathway, the world continues along with business as usual, largely in denial of the great dangers we face. Much of the world remains absorbed in a collective trance of materialism and consumerism. A majority of people continue within the mindset believing that we are separate from one another, from nature and from deeper belonging in the universe. Although diverse movements for transforming the society and restoring the ecology emerge, they are too small to penetrate through the distraction and denial of the majority. As a result, we fail to recognize and choose higher possibilities for regeneration and renewal.

Within a decade or so, delay and denial lead to calamitous breakdowns of every kind—ecological, social, financial, civic, psychological and spiritual. Planetary chaos ensues. Denial and confusion combine with psychological and institutional inertia to inhibit a timely response. Delay produces severe climate disruption that results in even greater social and economic chaos. Systems of all kinds—economic, civic, media, medical, educational—begin to breakdown and collapse. Many people focus on the survival of their immediate family and friends. A narrow scope of concern fosters divisions and conflicts. Humanity becomes hopelessly fragmented and dispirited. Failure to reconcile differences exacerbates violent conflict and produces deepening chaos. The Earth becomes so devastated and resources so depleted that the human species is unable to rise to a path of health and well-being. A new dark age stretches out before us.

Pathway 2: Authoritarian control with artificial intelligence. In this pathway, the dangers of climate disruption are recognized and, to reign them in, humanity trades personal freedoms and human rights for the safety promised by highly authoritarian societies. Digital dictatorships employ powerful computer technologies, integrated across multiple areas (financial, social, medical, educational, employment, etc.) to tightly control their massive population.

An often-cited example is now developing in China, which is determinedly creating a digital dictatorship using "social credit" scores combined with facial recognition systems and other technologies to monitor and control every person with an array of punishments and rewards.[37] Each person's cell phone and internet access are being assigned a unique number so they can be tracked. Transgressions that reduce one's public trust score range from the minor (jaywalking, playing video games too long) to the major (promoting "fake news," "thinking infected by unhealthy thoughts," and criminal activity).

Punishments range from public shaming (having your name and image posted publicly) to restricted work opportunities, diminished access to educational opportunities for yourself or your children, limited access to quality medicine, reduced internet speeds, and much more. Rewards include better job possibilities, better travel options (a plane instead of a bus), discounts on energy bills, easier access to hotels, and even better matches on computer dating sites.

With artificial intelligence accelerating, the combination of punishments and rewards for each individual produces a highly regulated and regimented society. Public opinion and discourse are controlled by banning topics from news sources, promoting "pro-social themes," extensive monitoring of internet conversations, restricting personal gatherings, and more. The result is a carefully watched, scrutinized and controlled society that lives within ecological limits but at the cost of many freedoms.

China is not the only country moving toward digital authoritarianism. Their "Great Firewall" approach to the internet appears to be spreading to other countries, including India, Thailand, Vietnam, Russia, Iran, Ethiopia, and Zambia.[38]

In this pathway, a highly controlled and regimented world emerges as people trade their freedoms for the security of staying alive. The world comes close to a devastating crash but is able to avoid that outcome when severe restrictions are placed on nearly every aspect of life, stopping the downward descent. The unraveling of the old world is brought to an abrupt halt as trends are placed under strict control and curl back upon themselves, just short of hitting an ecological wall. Numerous grassroots movements make determined efforts to realize transformational changes but are overwhelmed by and absorbed into the Authoritarian society that steadily takes hold. A future of constraint and conformity lies ahead.

Pathway 3: Great Transition and transformation. A pathway of regeneration and renewal emerges as humanity chooses to live within the ecological boundaries of the Earth and with conscious regard for the well-being of all life. This pathway does not hit rock bottom and crash; instead, it goes into the depths of tragedy and sorrow while narrowly avoiding a full crash.

The destructive unraveling of the old world meets constructive forces that are weaving together a new world. In the meeting, there is a turbulent period of transition as the evolutionary momentum of the past is gathered together into a new dynamic of mature regeneration. On the surface, this appears to be a time of confusion and chaos; yet, deep currents of transformation are at work, weaving the world together with a higher level of coherence, potential and purpose. Transformational trends provide enough forward momentum and creativity to reconfigure and uplift unraveling trends onto a new pathway of possibility.

This pathway does not produce an instant, new "golden age." Because the route forward emerges from immensely destructive forces, it requires a long stretch of healing and renewal for humanity to realize our potentials. Although this path of transformation is the most demanding of humanity's maturity and consciousness, it is a future that is within our capacities to choose as an increasingly self-aware and self-reflective species.

Choosing Earth focuses on the third pathway of transition and transformation as it is most life-affirming. Nonetheless, it is important to emphasize that the future is so open that no single pathway seems likely to dominate in the decades just ahead. Instead, the key issue is, "In what direction will the center of social gravity shift in the next half-century?" Which of these three pathways will become most influential for orienting our journey into the future?

The following graphic offers a visual representation of each of these three pathways. The descriptions of each stage ("the great unraveling," "the great fall," "the great sorrow," and so on) will become apparent as we work our way through the half-century scenario of great transition.

Each one of these three futures could emerge from the same set of driving trends now present in the world.

> *What distinguishes these futures is not*
> *the underlying trends but the overlying choices*
> *that we humans make for the future.*

There is no single, most likely future. The pathway that prevails will depend on what we consciously choose, or on that to which we unconsciously surrender. Because all three futures are plausible and possible, the most critical factors are the choices we make as a human community and the mindset or paradigm that guides those choices. Therefore, *a pathway of "Great Transition" is not a prediction*; instead, it is a plausible description leading to our early adulthood as a regenerative, planetary civilization.

> *Above all, the Great Transition scenario*
> *is an exercise in social imagination.*

One of our most important capacities as a species is our ability to look ahead consciously, anticipate what may unfold, and respond swiftly. If we can visualize in our social imagination how a ruinous or oppressive future lies ahead, then we don't have to enact that

Figure 2: Three Pathways Ahead

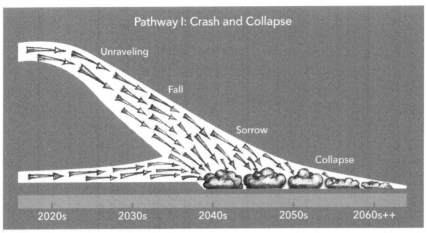

Pathway I: Crash and Collapse

Unraveling

Fall

Sorrow

Collapse

| 2020s | 2030s | 2040s | 2050s | 2060s++ |

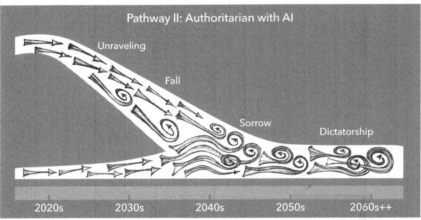

Pathway II: Authoritarian with AI

Unraveling

Fall

Sorrow

Dictatorship

| 2020s | 2030s | 2040s | 2050s | 2060s++ |

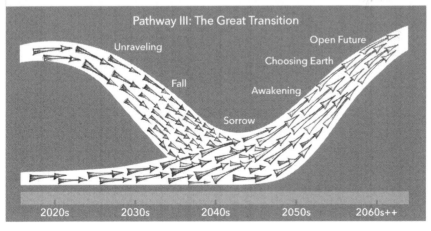

Pathway III: The Great Transition

Unraveling

Open Future

Choosing Earth

Fall

Awakening

Sorrow

| 2020s | 2030s | 2040s | 2050s | 2060s++ |

future in physical reality. If we can use our collective imagination to envision how we are creating an uninhabitable Earth, then we don't have to manifest that future in physical reality to learn its lessons. We can internalize the teachings and insights of the imagined future and act consciously to choose a different pathway ahead.

However, the reality of human evolution seems to be that we don't move and change in a dramatic way until we are forced by painful circumstances to do so. Therefore, a Great Transition pathway or scenario does not assume we humans will suddenly wake up and choose a regenerative future. Although it is conceivable that we could, this scenario assumes we will collectively choose and change only when driven to act by growing catastrophe and drawn ahead by consciously understood opportunity.

To some, the Great Transition scenario seems overly pessimistic while to others it is unduly optimistic. Whichever your view, its greatest usefulness is in awakening our collective imagination to visualize and reflect on what lies ahead. If we can recognize more clearly what lies ahead, we can consciously orient toward a different future—NOW.

With greater insight regarding both peril and opportunity, we can choose a more regenerative and thriving future. So, dear reader, please approach this alarming, shocking and yet realistic scenario as an exercise in social imagination intended to awaken our capacity for social dialogue and collective leadership for a different possibility, NOW, rather than after additional years of delay.

No matter how grim the future may look, it is important to recognize the many areas where we humans have long been working together successfully.

- **Weather**—The world weather system merges information from more than 100 countries every day to provide weather information globally.

- **Health**—Nations around the globe have cooperated to eradicate diseases such as smallpox, polio, and diphtheria.

- **Travel**—International aviation agreements assure the smooth functioning of global air transport while global cooperation has enabled the International Space Station to be built by a consortium of nations.

- **Communications**—The International Telecommunications Union (ITU) allocates the electromagnetic spectrum so that television signals, cellular phones, and radio signals are not overwhelmed with noise.
- **Justice**—Global ethics are emerging as world courts and tribunals hold heads of state accountable for policies of genocide, torture, and crimes against humanity.
- **Environment**—Despite lagging climate action, nations of the world have reached important agreements on ecological concerns such as banning CFC's that damage the atmosphere's ozone layer.

These examples of successful collaboration among the human community provide an important context for looking ahead—they illustrate humanity's capacity to rise to a higher maturity and work together when it really counts.

PART III:

A Pathway of Great Transition

6. Introduction to Great Transition

Real change involves real people with diverse lives directly engaging the world. In that encounter, great suffering becomes the evolutionary fire that burns through our attachments to old identities, dogmas and worldviews to awaken compassion for one another, and the rest of life. In that spirit, here is a short summary of my view of our time of great transition, taken from my 2009 book, *The Living Universe*:

> "The suffering, distress, and anguish of these times will become a purifying fire that burns through ancient prejudices and hostilities to cleanse the soul of our species. I expect no single, golden moment of reconciliation to descend upon the planet; instead, waves of ecological calamity will reinforce periods of economic crisis, and both will be amplified by massive waves of civil unrest. Instead of a single crescendo of crisis and conflict, there will likely be momentary reconciliation followed by disintegration, and then new reconciliation. In giving birth to a sustainable world civilization, humanity will probably move back and forth through cycles of contraction and relaxation. Only when we utterly exhaust ourselves will we burn through the barriers that separate us from our wholeness as a human family. Eventually we will see that we have an unyielding choice between a badly injured (or even stillborn) planetary civilization and the birth of a bruised but relatively healthy human family and biosphere. In seeing and accepting responsibility for this inescapable choice, we will work to discover a common sense of reality, identity, and social purpose. Finding this new common sense will be an extremely demanding task. Only after we have exhausted all hope of partial solutions will we be willing to move forward with an open mind and heart toward a future of mutually supportive development. Ultimately, in moving through our initiation, we can grow from our adolescent ways as a species into our early adulthood and consciously take responsibility for our relationship with the Earth, the rest of life, and the universe."[39]

Although this passage offers a summary view of our time of great transition, it does not describe in any detail the changes involved. To encourage a more robust conversation regarding our future, presented in this section is a scenario for the next 50 years—from 2020 to 2070—describing how the next half century could unfold if we were to make a collective transition to a genuinely sustainable and purposeful future.

Let's begin with a summary portrayal of the next half-century as a way to imagine the extraordinary arc of change involved in humanity's great transition to a mature planetary civilization.

Summary of Stages of Great Transition: 2020-2070

- **2020s: Recognizing Crisis.** *The Great Unraveling*—This is a decade of widespread awakening to the immediacy, scope and depth of the Great Transition. The world is coming apart at the seams. Economic supply chains are being broken. The world economy is decoupling and decentralizing. Breakdowns of institutions are widespread with the disruption of services of all kinds. Political and social chaos is growing. Trust and legitimacy of leadership is declining. Policies lack essential coordination, speed and depth. Despite growing recognition of a whole-systems crisis, many individuals and institutions resist making essential changes which exacerbates the Great Unraveling.

- **2030s: Collapsing Civilizations.** *The Great Fall*—Climate disruption is the catalyst for a full-blown systems crisis, producing a global financial crisis that fully unravels the world economy and society. A coherent, global response is too little and too late to meet rapidly changing conditions. Institutions are overwhelmed with the speed of the world coming apart and the scope of breakdowns and cascading collapse. Chaos and confusion are mounting along with deep social panic and anarchy. The world passes critical tipping points and with a Great Fall, drops into deep disorder and disarray.

- **2040s: Into the Fire of Initiation.** *The Great Sorrow*—As the consequences of climate chaos, financial breakdowns, civic anarchy, species extinction, mass migrations and widespread famines continue to grow, the entire world descends into collapse. It is a time of Great Sorrow and mourning for all that is being lost. The need for a profound transformation is anchored clearly in the raw experience of humanity. We recognize finally that we will either pull together in common effort or face the functional extinction of our species. We also recognize we may have waited too long and our efforts may be in vain.

- **2050s: A Conscious Species-Civilization.** *The Great Awakening*—Pushed to extreme limits by catastrophe, the suffering and sorrow of these times become a purifying fire that burns through remaining barriers to allow a Great Awakening of the collective mind and soul of our species. In turn, a new pathway emerges more clearly for humanity. We collectively acknowledge the need for a much more conscious species-civilization, grow into our early adulthood, work for broad and deep reconciliation, establish full-hearted communication, cultivate caring communities, and choose lifeways of sustainable simplicity.

- **2060s: Preserving the Future.** *Choosing Earth*—We leave behind the world of the past and turn toward new pathways ahead. There is wide recognition it will take centuries before the Earth's natural systems can recover. Nonetheless, with a new maturity and new consciousness, humanity is Choosing Earth and caring for the well-being of all life. After experiencing the reality of widespread collapse, many are growing into more conscious unity with one another and all of life. Although collapse and authoritarianism still remain as potential outcomes, the center of social gravity has shifted toward a healing journey for the Earth. The people of the Earth are working to establish an enduring, species-civilization for the long-term future.

- **2070s and Centuries Beyond. Three Pathways Ahead.** *An Open Future*—We have chosen a new relationship with the Earth as a foundation for the future. Three pathways still remain but the center of social gravity has shifted in favor of a transformational path forward. The future has not been closed by either complete ruin of the biosphere or by imprisonment in authoritarian structures. Instead, we have an Open Future and chance to realize a higher humanity and maturity as an enduring planetary civilization. A healing journey begins for the ecology of life on Earth.

The portrayal of the future that follows is based upon the best scientific estimates available, decade by decade, for the next 50 years across a number of key areas. These are:

1. Global warming and climate disruption
2. Water scarcity
3. Food scarcity
4. Climate refugees
5. Species extinction
6. World population
7. Economic growth/breakdown
8. Economic inequities

Where a number of studies focus on a few of these driving trends, my intention is to consider all eight and explore how they seem likely to interact and unfold over decades—both individually and as an interdependent world system.

Because the availability of "hard" trend data diminishes rapidly as we look further into the future, the early decades—especially the 2020s—are more heavily weighted with scientific data and analyses. Building from the foundation established by these scientific trends, a softer scenario or narrative is developed for each decade.

The Great Transition scenario begins with the same driving trends found in the pathways of Collapse and Authoritarianism. The difference is not in the early trends but in the choices the human community makes in response to those trends. A Great Transition emerges because we raise our heads and awaken our hearts to follow a higher purpose and possibility as a species.

6.1 The 2020s: Recognizing Crisis—The Great Unraveling

In this decade, the Great Transition gets underway in earnest as humanity gradually awakens to the unyielding fact that we are confronting a profound, world crisis. In stages we recognize that instead of a single problem to be fixed, we instead confront a whole systems crisis that requires deep changes in how we live on Earth. The people of Earth do not come to this understanding quickly or easily. Humanity enters this pivotal decade with deep divisions: Although concerns about climate change grew significantly prior to the 2020s, a substantial minority does not view this as a threat to human existence.[40] Others are distracted from the climate emergency by day-to-day concerns—ranging from survival pressures for the poor to the distractions of maintaining the status quo for those more wealthy. People with more education tend to be more concerned about global warming and, generally, women are more likely than men to be more alarmed about climate change.[41]

Slowly, the world awakens to the reality that we face a whole systems crisis that goes far beyond climate disruption and reaches deep into humanity's collective psyche and soul. Here are the major trends in this decade:

Major Driving Trends in the 2020s

- **Global Warming:** A 1.2° Celsius of global warming (roughly 2.2° Fahrenheit) by 2020 is producing clear evidence that major climate disruption is underway. Scientists are concerned that a 1.5°C increase will produce much more climate instability than previously thought.[42] There is growing alarm as scientific projections estimate a catastrophic temperature rise of 4° Celsius (7.2° Fahrenheit) will occur by the end of the century.[43]

  The implications for global warming are horrendous: For example, the September 2019 IPCC "Special Report on the Ocean and Cryosphere in a Changing Climate" recognized that half of the world's megacities with almost 2 billion people are located on vulnerable coasts. Even if global temperature rise is restricted to just 2°C, scientists expect the impact of sea level rise to cause several trillion dollars of damage a year and result in many millions of people migrating away from coastal areas.[44] The special report offers a grim picture of the long-term future:

Figure 3: Impacts of Global Warming by 2100[45]

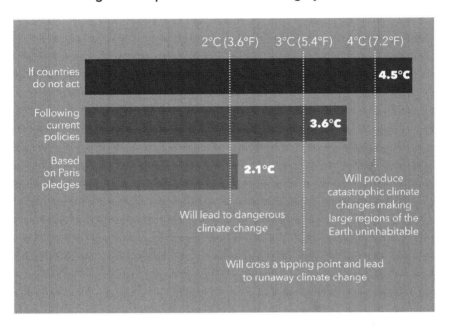

"We have simply waited too long to reduce emissions, and will be forced to grapple with impacts that can no longer be avoided. However, the difference between sharply reducing emissions and continuing along the 'business as usual' pathway is stark: Under a low-emissions scenario, managing the impacts of climate change will be expensive but possible; doing nothing will result in unmanageably catastrophic effects."[46]

Sea level rise will continue for hundreds and even thousands of years even if emissions are reduced to zero now.[47] Despite clear warnings of catastrophe, CO_2 emissions are continuing to grow.[48] This raises fears that we may create a 'hothouse Earth' condition that is unlike anything in human experience.[49]

In addition to temperature rise producing warming oceans, shrinking ice sheets, and ocean acidification, global warming is also bringing new weather extremes with storms, rain, floods, and droughts that severely impact agriculture and habitats.[50] All of these changes are expected to intensify throughout the 21st century and beyond.

Global warming has a direct impact on human health as well. A report from the World Health Organization states: "The climate crisis is a health crisis. . . that exacerbates malnutrition and fuels the spread of infectious diseases such as malaria. The same emissions that cause global warming are responsible for more than one-quarter of deaths from heart attack, stroke, lung cancer, and chronic respiratory disease."[51]

For a number of reasons, pandemics—diseases that spread worldwide—are more likely to emerge from conditions produced by global warming.

- First, as frozen regions of the Earth begin to thaw with global warming, they release viruses that have been locked away for tens of thousands of years. During the preceding ice ages, both humans and animals may have lowered their disease resistance and become much more vulnerable to their infections.

- Second, pandemics are more likely to emerge when economic advances support dramatic population growth and lead to large human populations living in close proximity with animals, enabling diseases to jump more easily from animals to humans.

- Third, with technological advances and high mobility, there is a more rapid mixing of animals and people around the Earth, giving viruses a way of making their journey around the world more swiftly. The scope and speed of modern human travel makes quarantines much more difficult—perhaps nearly impossible—to implement and enforce.

- Fourth, technological advances create the possibility of terrorists manufacturing or bio-engineering pathogens as bio-weapons to produce pandemic threats.

Pandemics—such as the coronavirus—have the potential to become a recurring disturbance in our rapidly warming world.[52] While pandemics are unlikely to be the catalyst for a global, civilizational collapse, they do reveal the vulnerability of our tightly interconnected social and economic systems. They also offer a compelling example of the need for mature, planetary collaboration. A swift global response shows that, once people recognize a genuine threat, governments can quickly organize and act, and people can change their behaviors. However, there are key differences between the climate crisis and pandemics. Although pandemics reveal we are all connected in the Earth's web of life, they are generally perceived as a relatively discrete, nearby, immediate and personal threat to one's self and family. Decisive action can protect people from many of these threats and this provides a formula for a broad response. In comparison, climate disruption is a more complex, distant, vague, and a general threat to society and the larger economy. Actions required are not simple and the benefits of those actions are less certain and less immediate. Ambiguity and uncertainty make a unified response and decisive climate action much more difficult.

Despite these differences, the coronavirus pandemic (just over one month old at the time of this writing) has the potential to disrupt business as usual and create an evolutionary tipping point for all of humanity. Quarantines, sickness and death may so interrupt and unsettle established patterns of living that we are pulled out of the habits, busyness and distractions of everyday life and given the social space to awaken to the rapidly developing, world systems crisis. The quarantine conditions and life-threatening situations created by the coronavirus could become an evolutionary wild card that shocks humanity into collective action with clear-headed choices for our common future.

- **Water Scarcity:** Although huge oceans cover the Earth, only 3% of the planet's water is fresh, and much of that is inaccessible—with over 2/3 of the fresh water locked up in icecaps and glaciers, and with nearly all the rest found in ground water. Only 3/10 of one percent of all fresh water in the world is found in surface lakes and rivers.

Given the enormous increase in world population and water-intensive ways of living, water has already become a scarce resource. In 2020, an estimated 30% to 40% of the world will experience water scarcity and by 2025, an estimated 3 billion people will live in areas plagued by water scarcity, with two-thirds of the world's population living in water-stressed regions.[53] In 2019, "844 million people, 1 in 9, lacked access to safe water and 2.3 billion people, 1 in 3, lacked access to a toilet."[54]

Over 2 billion people live in countries experiencing high water stress, and about 4 billion people experience severe water scarcity during at least one month of the year. Stress levels will continue to increase as demand for water grows and the effects of global warming intensify.[55]

- **Food Scarcity:** "In 2019, a little over 800 million people suffered from hunger, corresponding to about one in every nine people in the world."[56] Despite significant improvements in previous decades, the prospect for the future is grim due to climate disruptions.[57] To illustrate the predicament, "According to UNICEF, 22,000 children die each day due to poverty. And they die quietly in some of the poorest villages on earth, far removed from the scrutiny and the conscience of the world."

Around 27% of all children in developing countries are estimated to be underweight or stunted."[58] Global demand for food will more than double over the coming half-century as we add roughly another 3 to 4 billion people. A central issue in the coming half century is whether humanity can achieve and sustain such an enormous increase in food production.[59] Another study found that:

"[D]ecisions made in the next few decades will have huge ramifications for the future of our planet, and getting our food systems right is at the heart of this. Current practices are contributing to the problem, all in an effort to produce the record amounts of food needed to feed our global population. . . . it was this very progress that contributed to large-scale land and water degradation, biodiversity losses and increased greenhouse gas emissions. Now, the productivity of 23 per cent of global land has declined, and about 75% of freshwater is used just for agriculture."[60]

- **Climate Refugees:** Between 2008 and 2015, an average of 26.4 million people per year were displaced by climate or weather-related disasters, according to the United Nations.[61] Tens of millions of people are expected to be on the move in 2020.

- **Species Extinction:** By the end of this century, a UN Report concludes that more than one million species of plants and animals are at risk of extinction—many of which are predicted to be pushed into extinction within just a few decades. Robert Watson, a British chemist who served as the panel's chairman, stated, "The decline in biodiversity is eroding the foundations of our economies, livelihoods, food security, health and quality of life worldwide."[62] The integrity of the biosphere is being devastated and losses include insects, birds, mammals and reptiles as well as fish. The overall outlook is very grim.

The world's **insects** are hurtling down the path to extinction, threatening a "catastrophic collapse of nature's ecosystems" according to the first global scientific review.[63] The analysis found that more than 40% of insect species are declining and a third are endangered. The rate of extinction for insects is eight times faster than that of mammals, birds and reptiles and is so great that, "Unless we change our ways of producing food, insects as a whole will go down the path of extinction in a few decades. . . . The repercussions for the planet's ecosystems are catastrophic to say the least."

Bees are also disappearing at an alarming rate due to the excessive use of pesticides in crops and the spread of certain parasites that only reproduce in bee colonies. *The extinction of bees could mean the end of humanity.* If bees didn't exist, it is hard to imagine humans surviving. Out of the 100 crop species that provide 90% of our food, 35% are pollinated by bees, birds and bats.[64]

Another study found that birds are vanishing from North America: The number of **birds** in the United States and Canada has declined by 3 billion, or 29%, over the past half-century.[65] David Yarnold, president of the National Audubon Society, called the findings "a full-blown crisis." Kevin Gaston, a conservation biologist, said that new findings signal something larger at work: "This is the loss of nature." "The skies are emptying out. There are 2.9 billion fewer birds taking wing now than there were 50 years ago."[66] The analysis, published in the journal Science, is the most exhaustive and ambitious attempt yet to learn what is happening to avian populations. The results have shocked researchers and conservation organizations.

The **ocean's** eco-system is being devastated with marine life declining by 49% between 1970 and 2012. Overfishing and pollution are producing an "unprecedented" marine extinction. A major report found that every species of wild-caught seafood—from tuna to sardines—will collapse by the year 2050. "Collapse" was defined as a 90% depletion of the species baseline abundance.[67] Another report warns that hunting and killing of the ocean's largest species will disrupt ecosystems for millions of years.[68]

Here is how the Center for Biological Diversity describes the extinction crisis:

"Wildlife populations are crashing around the world. . . . Our planet now faces a global extinction crisis never witnessed by humankind. Scientists predict that more than 1 million species are on track for extinction in the coming decades. Wildlife populations around the world are crashing at alarming rates and with distressing frequency. . . . When a species goes extinct, the world around us unravels a bit. The consequences are profound, not just in those places and for those species but for all of us. These are tangible consequential losses, such as crop pollination and water purification, but also spiritual and cultural ones. Although often obscured by the noise and rush of modern life, people retain deep emotional connections to the wild world. Wildlife and plants have inspired our histories, mythologies, languages and how we view the world. The presence of wildlife brings joy and enriches us all—and each extinction makes our home a lonelier and colder place for us and future generations. The current extinction crisis is entirely of our own making."[69]

- **World Population:** At the beginning of the 2020s, world population is roughly 7.8 billion.[70] Although population projections to the end of the century are difficult, a median estimate of total world population in 2100 is approximately 11 billion. Rough estimates are that, by 2100, the top five most populous countries will be: India, with 1.5 billion people, China with 1 billion, Nigeria with nearly 800 million (comparable to the entire population of Europe in 2010), the US with 450 million, and Pakistan with 350 million.[71]

Figure 4: World Population Growth: 1750–2100[72]

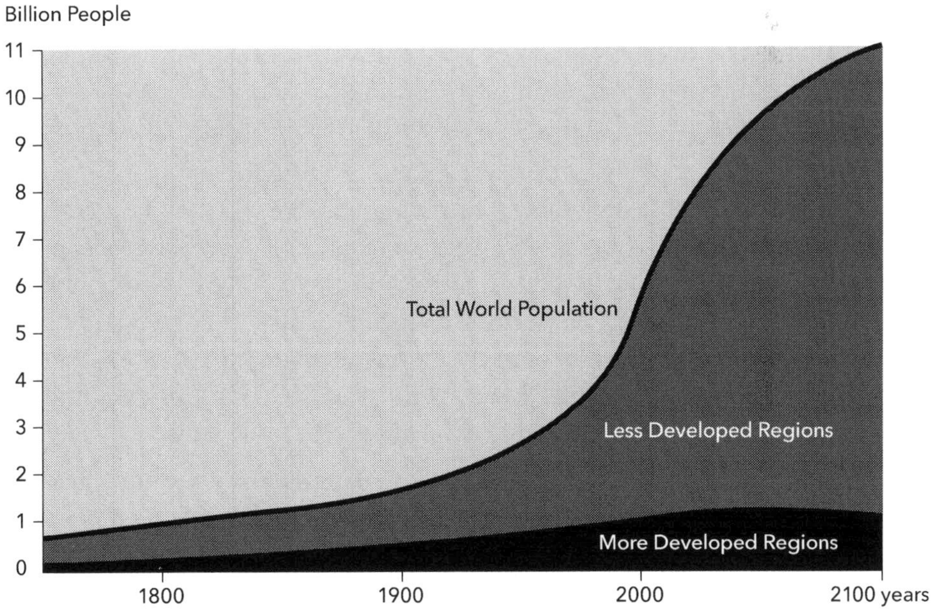

Billion People

Less developed regions: Africa, Asia (excluding Japan), Latin American and the Caribbean, and Oceana (excluding Australian and New Zealand).

More developed regions: Europe, North America (Canada and the United States), Japan, Australia, and New Zealand.

An estimate of global population of roughly 11 billion is far from certain—particularly if deep and rapid shifts to sustainable ways of living are not adopted. Given current food production capacity and water resources, the Earth can support roughly 9 billion people *if resources are shared equally*. However, with agricultural productivity falling due to global warming and water scarcity, the carrying capacity of the Earth is declining.

In addition, much depends on the consumption patterns of developed nations relative to the rest of the world. If the entire world consumed at the same level as the United States, the Earth could support roughly 1.5 billion people. With middle class European lifestyles, the carrying capacity grows to roughly 2 billion people.[73] The Earth supports US levels of consumption only because we are drawing down the "savings account" of non-renewable resources, such as fertile topsoil, clean drinkable water, virgin forests, undiminished fisheries, and untapped petroleum.

Our "savings account" is already running low and we are obliged to live within our means as a species. In turn, the carrying capacity of the Earth depends not only on the number of people on the planet, but also their levels and patterns of consumption. In the early 2020s, the human community is consuming the Earth's renewable resources at one and three-quarters times the sustainable rate[74]—and that is with roughly 6 billion people involuntarily living in "low carbon lifestyles" and consuming next to nothing compared to the American middle class.

Given the great reluctance of wealthier nations to sacrifice their high-consumption lifestyles, and given that the consumption footprint of the Earth is rapidly approaching nearly double what the Earth can provide on a long-term basis, it seems likely that a die-off in human numbers will occur. Is the great suffering that will result required to push people in developed nations to make needed changes in their levels and patterns of consumption? How much pain and suffering are required for humanity to turn to a new equilibrium and fairness in global consumption?

- **Economic Growth/Breakdown:** Experts widely agree that roughly 70% of economic activity in the US is connected with the production of consumer goods, which is understandable for a consumer-based economy.[75] Numerous studies conclude, "Emissions are a symptom of consumption and unless we reduce consumption we'll not reduce emissions."[76] Therefore, future economic growth likely will be diminished by the urgent need to reduce carbon emissions and therefore the need to reduce overall consumption levels. "It doesn't matter if you're in a hot or a cold climate, a rich country or a poorer one—unchecked climate change is going to devastate the economy. This research comes as the United Nations says that climate impacts are happening faster, and hitting harder, than anticipated."[77] The risks associated with climate change are not being integrated into pricing and this, in turn, is reducing incentives needed to decrease emissions—an economic error with catastrophic consequences.

  "The next two decades will be decisive. They will determine whether we suffer severe and irreversible damage to livelihoods and the natural world or whether, instead, we set off on a more attractive path of sustainable and inclusive economic development and growth. . . If we go on emitting greenhouse gases at current rates for the next two decades, then it is likely that we will far exceed a 3°C increase. . . A rise of 3°C would be extremely dangerous, taking us to a temperature we have not seen on this planet for around 3 million years. . . A warming of this magnitude could transform where we could live, severely damage livelihoods, displace billions of people and lead to severe and extended conflict."[78]

- **Economic Inequities:** It doesn't matter how we look at it—global inequality of wealth and income is getting worse—much worse. In 2017, the world's six richest men were as wealthy as half of humanity![79] Six individuals have as much wealth as 3,600,000,000 of the world's poorest people. Equally stunning is the estimate that the richest 1% of the world's population has more wealth than the rest of the world's population combined.[80]

The astonishing inequity in the United States is revealed by the astounding fact that tax rates for the wealthiest Americans are lower than for any other income group: "For the first time on record, the 400 wealthiest Americans last year paid a lower total tax rate—spanning federal, state, and local taxes—than any other income group."[81] As long as a wealthy elite has the power to set the rules to their own advantage, inequality will continue to worsen.[82]

A powerful way to visually represent the unfairness and injustice of global income distribution is by looking at the shape of the following figure where income for the world is divided into five groups, each representing 20% of the world from low to high income. The long, thin portion of the outline (akin to the stem of a champagne glass) represents the annual income of a majority—approximately 60% of the people in the world. The portion where the stem begins to widen out appreciably represents the income of the next 20%—the global middle-class. The widest portion illustrates the income received by the world's wealthiest 20%. Just by looking, it is apparent that the human family is made up of a huge impoverished class, a small but growing middle class, and a very small and very wealthy elite.

These inequities have major consequences for the Earth's climate. Almost 50% of global carbon emissions are generated by the activities of the wealthiest 10% of the global population. In stark contrast, the poorest 50% of the global population are responsible for only around 10% of global carbon emissions, yet live overwhelmingly in the countries most vulnerable to climate change.[83] Given these immense disparities, climate adaptation is already a profound issue of social justice

A fundamental proposition of climate justice is that those who are least responsible for climate change suffer its gravest consequences.[84] Structural inequities based on race mean that communities of color will continue to be hit "first and worst" in the climate crisis.[85] Yet, by imposing a per-capita carbon emissions limit on the top 10% of global emitters (roughly equivalent to that of an average European citizen), global emissions could be reduced by one-third in a matter of a year or two.

Figure 5: Global Income Distribution[86]

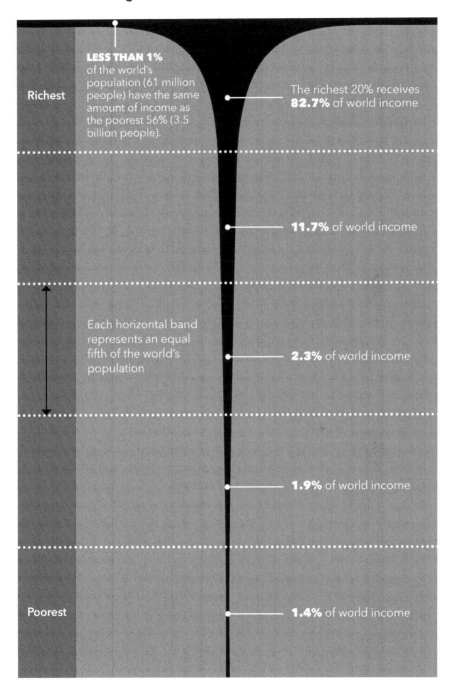

LESS THAN 1% of the world's population (61 million people) have the same amount of income as the poorest 56% (3.5 billion people).

Richest

The richest 20% receives **82.7%** of world income

11.7% of world income

Each horizontal band represents an equal fifth of the world's population

2.3% of world income

1.9% of world income

Poorest

1.4% of world income

Because great disparities of wealth have been a consistent precursor to dramatic social change, humanity is being challenged to recognize the current economy is not working for the benefit of the majority. A voluntary change in favor of a much more equitable distribution of wealth would be a very wise course of action.

Scenario: How the Decade of the 2020s Unfolds

In this decade, the human community begins to recognize that global warming is changing the world in such profound ways that life will never be the same. Keeping long-term warming of the planet below the target of 1.5° Centigrade (or 2.7° Fahrenheit)—the goal set in the Paris Climate agreements signed in 2015—seems impossible as it requires immediate and dramatic cutbacks in CO_2 emissions which in turn require radical changes in the energy uses and lifestyles that are producing these emissions.

The Paris agreements also include pathways for developed nations to assist developing nations in their climate mitigation and creative adaptation efforts.[87] Yet, at the start of this pivotal decade, CO_2 emissions are increasing and attempts to reduce them through coordinated actions among nations have failed. Global CO_2 emissions are on track to produce a dangerous, 2° Centigrade (3.6° Fahrenheit) increase in temperature as soon as the end of the decade.

At the outset of the 2020s, many people are unaware and ill-informed regarding how profoundly global warming will impact the future of life on the planet. As people learn how serious our situation has already become, responses range from denial and disbelief to confusion and alarm as people confront the need to adapt to a reality beyond anything in prior human experience. Wealthy elites who dominate business, politics, media, and finance regard global warming, species extinction, and other trends as important but exaggerated. Most leaders are among a privileged minority who are immersed in the comforts of wealth, status, pleasure and power, and are distracted by the busyness and demands of everyday life. Their primary concern is with continuing business as usual despite growing alarm among scientists and protests by youth, academics, women, and the general public. Instead of mobilizing for dramatic action and innovation, privileged elites seek only gradual adjustments that don't disrupt the status quo.

Mainstream media strongly support the social trance of consumerism with endless entertainment—sports, reality television, movies, video games, and celebrity gossip—that glamorizes materialist lifestyles and deflects and numbs social attention.

Although climate disruptions and a cascade of other difficulties are growing visibly, influential leaders soften claims of an intertwined web of problems. Instead, these problems are portrayed as:

- Not as important as other issues such as jobs and healthcare.
- Not as urgent or immediate as claimed, so, there is ample time to respond.
- Not as large in scope as claimed; rather, these are pockets of problems to be tackled.
- Not as difficult to remedy as claimed; technology will fix many of the problems.
- Not as a whole-systems crisis; rather, these are individual problems that can be solved one at a time.
- Not a graspable problem I can solve: "What can I do? I'm only one person."
- Not my responsibility: "I did not create this mess, so why ask me to clean it up?"

The "soft denial" of many leaders combines with a pervasive sense of helplessness. Understandably, business as usual persists and mainstream institutions respond with half-hearted measures that do little to slow a relentless advance toward a disastrous future. Yet, small but growing communities of people are shifting toward adaptations, seeing the challenges we face as an urgent, whole systems crisis.

The United States—the world's leading consumer nation—illustrates the difficulty of dealing with transition in a healthy way. Reverend Victor Kazanjian of the United Religions Initiative describes how the US is a society of grievance that is unable to accept our fate and grieve over the changes that are required of us. He writes:

"... much of what underlies anger, rage, and violence is grief. A sense of loss upon loss upon loss. But in our culture, we don't have much space for grief. Grief when unaddressed becomes grievance. We are in a culture of grievance. Grief being expressed as blaming the other. We have to address deep grief."

Despite great resistance, by the middle of the 2020s, the disruptions in natural systems become so great that they begin to break the consensus trance of consumerism, distraction and denial. Climate emergencies multiply and awaken a growing recognition that Earth-scale challenges are underway. Complacency gives way to growing alarm as seasons around the planet are so disrupted that food production is compromised, producing regions of severe famine and civic unrest.

The overarching challenge of the 2020s is to awaken our social imagination to the imperative of making extraordinary changes in how we live on the Earth and to acknowledge that an entirely new pathway into the future is required if CO_2 emissions are to be reduced and brought under control.

- Gradually, those who are more materially privileged begin shifting from overconsumption toward lifestyles of "voluntary simplicity" while those who are impoverished continue with involuntary simplicity and a daily struggle for survival.

- For the more affluent, diets begin to shift toward vegetarianism, transportation shifts toward electric vehicles, homes become more energy efficient, and work shifts toward decreased environmental impact and increased social contribution and meaning.

- Ecological lifestyles move from a fringe movement for a few to a wave of experimentation for mainstream culture. Low carbon lifestyles that are materially simple and experientially rich become more widespread. For some it's a relatively superficial way of "going green" and, for others, it's a full reorganization of their way of living.

- Materialism and consumerism are increasingly called into question as people challenge cultures of aggressive advertising, declaring that we are more than consumers to be entertained; we are citizens of the Earth who want to participate in creating a more sustainable future.

- People begin to localize activities—ranging from experiments with pocket neighborhoods and community gardens to supporting zoning changes that allow development of "eco-village" structures of living.
- New configurations of economic activity begin to emerge that emphasize local resilience with new skill sets and patterns of work.

By the end of the decade, a transition in culture and consciousness is getting underway, primarily in rich countries where people have the luxury of focusing on more than daily survival. A clear-headed understanding is growing that new approaches to living are essential.

The growth of meditation and other transformational practices—such as yoga and psychological self-help—are supporting a shift in consciousness. Small but significant numbers of people are waking up and expanding their identity beyond a separate, biological sense of self. Others are consciously feeling and grieving the losses the Earth is experiencing—both as individuals and in group rituals. A growing portion of the population is laying the groundwork for humanity to wake up and grow up.

The urgent need to develop a new public understanding with working agreements for building a sustainable future promotes the rise of diverse communication movements. These movements range from living room conversations and civic dialogues to conferences among leaders in business, government, media, education, religion and more. Other experiments with social media emerge including trans-partisan "community voice" experiments that combine the internet with social media and television to create "Electronic Town Meetings" at the metropolitan scale. These diverse communication movements provide a vital source of social cohesion to hold the unraveling world together.

6.2 The 2030s: Collapsing Civilizations— The Great Fall

In the 2030s—the time many people are attempting to mobilize for action—the institutions on which we depend are breaking down. The pace of global warming is racing past the ability of aging institutions to keep up. Local and national governments, financial organizations, academic institutions, religious organizations and more are overwhelmed trying to understand what is happening and under-resourced in trying to respond.

Massive indebtedness created by extravagant spending in earlier decades now prevents many institutions from mobilizing resources for creative action. Instead of rising to challenges, many institutions are unraveling and coming apart. Bankruptcy spreads to entire cities. Many vital services falter—including police and fire protection and the upkeep of infrastructure such as roads and electrical grids. Large corporations go bankrupt, resulting in job losses for huge numbers of people. Major colleges and universities become insolvent and close their doors. Many big churches cannot afford the upkeep and fail.

Breakdowns spread in waves throughout the world and people must increasingly fend for themselves at the local level. Instead of creative action to avert a deepening climate crisis, the world is preoccupied in coping with rapidly spreading breakdowns. Here are key trends that are driving this decade of breakdown and stress:

Major Driving Trends in the 2030s

- **Global Warming and Climate Disruption:** Global temperatures increase over historical levels by 2° Celsius (3.6° Fahrenheit) by the end of the 2030s. With a 2°C increase, ice sheets begin irreversible disintegration that will produce a catastrophic sea level rise, most dramatically in the next century. In addition to producing warming oceans, shrinking ice sheets and ocean acidification, temperature rise is also bringing new extremes of storms, rain, floods, and droughts that severely impact agriculture and habitats.[88]

A 2°C increase is the threshold for a critical tipping point and the beginning for runaway climate change.[89] The potential for unstoppable warming begins with the release of the "sleeping giant" of methane which is roughly 20 times more potent as a greenhouse gas than CO_2.[90] A surge in atmospheric methane threatens to erase the anticipated gains of the Paris Climate Agreement.[91] A key risk is that self-reinforcing feedback loops will push the climate into chaos before we have time to restructure our energy system.

Another "sleeping giant" is the Amazon rainforest, which has been viewed as a CO_2 "sink" that absorbs carbon. However, a recent study shows that tropical forests are losing their ability to absorb carbon, which will turn the Amazon into a source of CO_2 by the 2030s and accelerate climate breakdown, producing much more severe impacts requiring a much faster reduction in carbon producing activities to counteract the loss of carbon sinks.[92]

- **Climate Refugees:** With climate disruption, the number of refugees climbs from tens of millions on the move to a hundred million or more migrating to more favorable areas by the end of the 2030s. Migrations of this magnitude overwhelm the capacity of regions to adapt. For perspective, roughly one million refugees destabilized much of Europe in the 2010 decade. With a hundred million or more migrating, the impact produced is many times greater and is spread unevenly, mostly throughout the more resource favored northern hemisphere.

- **Water Scarcity:** Global demand for water exceeds sustainable use by 40%.[93] By 2030, at least three billion people suffer from water shortages.[94] With growing drought conditions, major cities around the world are beginning to run out of water. In 2019 Cape Town, South Africa came close to "zero day"—the day when the city runs out of water. Cape Town is just the beginning. At least 11 other major cities are likely to run out of water in this time frame: São Paulo, Brazil; Bangalore, India; Beijing, China; Cairo, Egypt; Jakarta, Indonesia; Moscow, Russia; Mexico City, Mexico; London, England; Tokyo, Japan; and Miami, USA.[95]

"In India, a country of 1.3 billion people, fully half the population lives in a water crisis. More than 20 cities—Delhi, Bangalore, and Hyderabad among them—will gulp their entire aquifers dry within the next two years. This translates into a hundred million people living with zero groundwater."[96]

- **Food Scarcity:** For every degree Centigrade of temperature rise, a 10% to 15% decrease in agricultural yields is expected. Therefore, a 2°C (3.6°F) temperature increase is expected to reduce agricultural productivity by 20% to 30% at a time when demand already stretches food supplies to their limits. Pockets of food scarcity grow into areas of outright famine producing further mass migrations and civic breakdowns. Dramatic changes in the daily diets of the people of the Earth bring home the reality of climate disruption. With the stress of heat, changes in growing season and locations, a range of foods are becoming prohibitively expensive for all but the most affluent. See the list at right to explore how diets may be impacted.[97]

FOOD SCARCITY

In the coming decades, a range of foods will become prohibitively expensive for all but the most affluent. An illustrative list is below. It is enlightening to go down the list and check off those foods you will greatly miss as they become increasingly costly. Unless you are growing many of these yourself or have considerable wealth, these foods will be virtually unavailable. This is a visceral example of the climate crisis hitting home.

❏ Almonds	❏ Coffee	❏ Potatoes
❏ Apples	❏ Corn	❏ Squash
❏ Avocadoes	❏ Honey	❏ Rice
❏ Bananas	❏ Maple Syrup	❏ Shrimp
❏ Chicken	❏ Oysters	❏ Soybeans
❏ Chocolate (Cacao)	❏ Peaches	❏ Strawberries
❏ Cod	❏ Peanuts	❏ Wine (Grapes)

People are beginning to create new diets that adapt to reduced choices for basic foods. Poorer people are forced to accept diets with diminished nutrition, reduced variety and less flavor—a significant decline in well-being and quality of life. A food revolution is underway that privileges the wealthy who can buy their way out of food limitations with genetically modified, greenhouse produced foods at much higher costs.

- **World Population:** Human numbers are expected to reach 8.5 billion to 9 billion by 2037.[98] A world population of 9 billion at the end of the 2030s is a realistic estimate. Much of the growth is in Africa, India, and southern Asia.

- **Species Extinction:** Building upon projections made in the 2020s that estimate a million species could be extinct by the end of the century, the loss of animal and plant species is rapidly accelerating.[99] The integrity and health of the Earth's biosphere (plants, land animals, birds, insects, and ocean life) is deteriorating rapidly. With regard to oceans, oxygen loss driven by global warming (and nutrient pollution produced by runoffs from agriculture and sewage) is suffocating the oceans with far-reaching and complex biological implications and producing marked declines in ocean life.[100]

- **Economic Growth/Breakdown:** The global economy is in deep turmoil and crisis given the extraordinary demands for making an extremely rapid transition to renewable sources of energy. Overall growth has come to a standstill despite dramatic increases in renewable energy innovation. Tremendous economic and social pressures are shifting more developed nations away from a focus on unrestrained growth and consumerism.

Around the world, creative entrepreneurs are experimenting with practical ways to recreate the economy so it works for both people and the planet. A widely recognized goal is to create self-organizing, regenerative forms of economic activity that work for global civilization as a whole.[101] With massive displacement of workers through automation combined with dislocations through climate disruption and the breakdown of large-scale factories and corporations, regenerative approaches to living are fostering the development of "local living economies."

Regenerative economies nested within alternative forms of community are emerging around the world to create more resilient living systems. Nonetheless, the magnitude of changes required to make a global transition to renewable energy, accompanied by regenerative economies designed with fairness and equity, seem insurmountable.

- **Economic Inequity:** The richest 1% are on target to own two-thirds of all wealth by 2030.[102] Vast disparities in wealth coupled with the economic demands for shifting to a net zero-carbon economy by 2050 are placing extreme pressures on the already disrupted global economy and society. An extreme lack of fairness and trust drains legitimacy from the world's economic system.

With enormous disparities in wealth and incomes, this is a decade of cascading breakdowns where vulnerable areas of the world experience full economic collapse. The growth paradigm of materialism and consumerism is disintegrating as a compelling social goal—not only is this paradigm undermining the well-being of most people, it is simultaneously devastating the Earth's biosphere.

Scenario: How the Decade of the 2030s Unfolds

In the decade of the 2030s, most people around the world recognize that a climate catastrophe is unfolding. Yet, entrenched bureaucracies in business, media, education, religion, social services, and more are unprepared and ill-equipped to meet the challenges of a worsening climate, deteriorating economies, and a collapsing biosphere.

In more affluent countries, most people are deep in debt, taxes are profoundly unequal, and the engines of economic growth are faltering. There is a rapid turnover of leaders and policy solutions but nothing seems to work for long. Efforts to create order are overwhelmed by growing levels of disorder. Large-scale social cohesion is very low and many leaders govern with virtually no support.

Previous levels of resilience are being depleted in a grinding, downward spiral of bureaucratic confusion and chaos.[103] We no longer have the capacity to bounce back quickly from difficulties. Some people seek security by turning toward more controlled, authoritarian districts. Others turn toward self-organizing communities that depend on strong relationships and collaborative approaches to living.

As climate disruption deepens, divisions of every kind are growing—financial, political, generational, gender, racial and ethnic, religious, and more. The only constant of this disorienting and confusing decade is the unrelenting stress of deep separations. The world is so awash with so many disputes at so many levels with so many differences of so many kinds that little room exists for rising to a higher humanity. The world is filled with blame, fault-making, denunciation, hostility, condemnation, and reproach.

The wealthiest billion people who enjoy a "good life" of material comfort and advantage face a growing outcry of protest from the eight billion struggling for continued existence. Nonetheless, wealthy elites resist making rapid adaptations to new ways of living. Having invested their lives and identities in material accumulation, they fight back, claiming their privilege is earned and deserved. Although most recognize the new realities, many reject new norms for living. However, by the end of the 2030s, their efforts to separate into isolated communities begin to falter as billions of impoverished people, with nothing to lose and much to gain, rise in protest.

With breakdowns increasing, localization is growing and encouraging a rush of social, economic, and technical innovations. Pocket neighborhoods are growing into diverse forms of eco-villages, which are establishing a resilient foundation for transition towns and sustainable cities. Newly organized communities are building more than physical structures; they are developing a fresh understanding of human character and a maturity that seeks to serve the well-being of all. Work roles are changing dramatically as small, self-organizing communities provide new settings for developing diverse skill sets for living.

Pushed by the climate crisis and spreading breakdowns, the affluent majority in developed nations recognize we must transform cultures of consumerism and reduce our ecological footprint if we are to avoid catastrophe. The cultural hypnosis of materialism loses its potency as people recognize the dream of unrestrained consumption is producing a devastating future for the Earth. A global culture valuing simplicity and sustainability is emerging. Mass media advertising that has been aggressively promoting the trance of consumer culture is shifting from product commercials to "Earth commercials" as companies proclaim their commitment to a healthy planet.

Wealthy countries are responsible for climate change, but it's the poor who are suffering most. Given the asymmetry of a disproportionate impact of global warming on poorer countries, wealthier nations are pressed—with only modest success—to take responsibility for supporting adaptations. Yet, strong initiatives are vital to foster a sense of global unity and cooperation as climate change is increasingly impacting the basics of life in poorer nations—water availability, food production, health, environmental quality and the well-being of vulnerable populations, especially women and children.

In poorer countries, the impacts of global warming are reversing progress in gender equality as men are often forced to migrate to find work, leaving women to handle the entire burden of raising children, farming or fishing locally, and managing the household. This leaves women more isolated and less able to find meaningful work and education.

In recognizing the adverse impacts of global warming on developing nations, a global movement for compensation, reparation, and adaptation grows, seeking to build a new sense of partnership among the people of the Earth.

Trans-partisan, "Community Voice" movements are continuing to grow around the Earth and connecting humanity into ever larger communities of conversation. Recognizing that the scale of conversation must match the scale of challenges, "Earth Voice" conversations that use widely available cell phones and social media are becoming established in the world. Media activism is also growing as people recognize the extent to which the cultural trance of consumerism has been produced by an avalanche of advertising over the course of multiple generations.

Recognizing the mass media is our "social brain" and a direct expression of collective intelligence, people realize that as the media goes, so goes our future. Media institutions are being held accountable to an entirely new degree and are being mobilized to awaken social imagination and visualize pathways of progress toward a sustainable and meaningful future.

The old world is unraveling and, despite the imperative of local to global transition, we still lack the overall support needed to move swiftly into this new world. Consumer society and ways of living change slowly, the dispossessed continue to be largely ignored, a green transition is unable to mobilize a majority into dramatic action, and authoritarian districts continue to separate themselves into compartmentalized areas of control.

Given these deep divisions, the 2030s are a time of churning chaos and conflict without an overarching set of values and intentions for moving ahead. The struggle for a new paradigm of living is fully underway. People ask: How can we once again feel at home on the Earth? Do we have the collective maturity to consciously engage in deep adaptation and make a great transition to a new pathway for the future?

6.3 The 2040s: Into the Fire of Initiation— The Great Sorrow

In the decade of the 2040s, we recognize that we are losing the race with climate catastrophe. Runaway climate disruption is no longer a looming possibility—it is a present reality. In this decade, we move beyond 2°C (3.6°F) of warming and approach 3°C (or 5.4°F)—a critical, climate tipping point.[104] Methane is surging into the atmosphere and pushing feedback loops into motion.[105] The world is moving beyond breakdowns and toward collapse and climate catastrophe. The world is thrown off balance and into collective despair.

We are out of time.

For over two decades, the human community has recognized we will eventually reach a point of no return. That point is now a painful reality. We can never go back to the earlier world we knew. It is far too late to "fix" the unraveling world situation. Feelings of shame, guilt, grief and despair grow as a ruinous future looms. Humanity is awash with unrelenting sorrow as the Earth of the past is left behind, never to return.

Major Driving Trends of the 2040s

- **Global Warming and Climate Disruption:** Global warming moves decisively beyond 2°C and approaches 3°C (5.4° Fahrenheit) and into the range where permafrost melting and methane release produces self-reinforcing feedback loops with runaway climate change. An already turbulent and chaotic climate is further amplified with consequences mounting to catastrophic proportions. Climate extremes include both fire and water—large regions of the Earth are experiencing unprecedented drought that brings fires to scorch a drying Earth while other regions are experiencing unprecedented storms, floods and sea level rise.[106]

- **Water Scarcity:** Water shortages are critical for 3 billion and likely more persons. This scarcity produces a dramatic increase in the number of climate refugees fleeing drought-stricken regions.

- **Food Scarcity:** Growing population pressures combine with climate disruption, falling agricultural productivity, water scarcity, and great economic inequities to produce large areas of devastating famine.

- **Climate Refugees:** At least 200 million climate refugees are expected to be on the move, creating colossal social and economic disruptions as communities in resource favored areas try to cope with the influx of overwhelming numbers of people.

- **World Population:** Population continues to grow toward a mid-century estimate of 9 billion people. However, population growth runs up against increasingly severe limits created by water and food scarcity and the breakdown of ecosystems.[107]

Great suffering becomes the catalyst that awakens humanity to action. A tragic estimate is that 10% of the Earth's poorest and most vulnerable populations will be at great risk of dying during this time of great transition. With a global population of roughly 9 billion people in the 2040s, this estimate translates into an order of magnitude of roughly 900 million people who could perish before a transition is accomplished. These millions will not die quietly and out of sight but, in our media rich world, will die very publicly, painfully and visibly. Their deaths will be caused by famine and disease but also by enormous levels of violence in conflicts over resources.

The die-off of hundreds of millions of people will produce unimaginable levels of moral and psychological trauma for the people of Earth. The needless suffering and death of hundreds of millions of people may awaken the psyche and soul of humanity to choose a path of equality and fairness in how we live together on Earth.

- **Species Extinction:** Decades of destruction of ecosystems are undermining the foundations for life around the world. The unyielding reality of ecological collapse reveals that we are an integral part of the global web of life and the threat of extinction includes humans as well. Countless species are going extinct, leaving the Earth an ever more barren world.

- **Economic Growth/Breakdown:** Economic breakdowns are spreading around the world, producing the full-scale collapse of vulnerable economies. Although economic breakdowns slow greenhouse gas emissions, widespread efforts for survival have the unfortunate result of pushing people and communities to use whatever energy sources are readily available, including coal and oil for short-term survival. A return to fossil fuels contributes to greenhouse gas emissions at the very time it is vital they are reduced. Although efforts for a deep reconfiguration of the local to global economy are underway, collapsing economies and eco-systems are making these efforts incredibly difficult.

- **Economic Inequities:** The immensely complex and difficult transition to a renewable energy economy reduces output, and the world is more challenged than ever to meet the needs of the world's poor and to move toward much greater fairness.

Global tensions between the haves and have-nots are accelerating past breaking points. The global crisis of fairness and social justice conflicts with cultures of consumerism producing a fierce struggle for the future direction of our species.

People with the least access to resources face the greatest challenges in adapting to global warming—and this is true across race, gender, age, geography, and class differences.[108] Widespread efforts grow for producing the essentials of life inexpensively and cutting back on luxury living for the affluent. The redistribution of land is also a key factor in fairness and is awakening titanic struggles over ownership and sharing.

Scenario: How the Decade of the 2040s Unfolds

The great transition has become a time of great suffering beyond anything humans have ever experienced.[109] A global collapse is underway producing all kinds of shortages—vital medicines and medical care, basic foods, clean water and mass transit. Broken supply chains and shortages produce hoarding, looting, black markets and hyperinflation. Major corporations go bankrupt as well as major cities whose tax base has dissolved. Key infrastructure is abandoned and falls into disrepair as nearly all maintenance is neglected—including electrical and telephone utilities, internet services, roads, bridges, traffic lights, sewage systems, garbage removal, and water systems. Neglect of infrastructure produces countless local disasters.

Confusion, chaos and conflict grow. As lawlessness spreads, private protection forces replace traditional police and law enforcement. At a larger scale, collapse spreads beyond cities to states and even nations. As nations breakdown in bankruptcy and disassemble, so do international organizations such as the United Nations, which endure as little more than symbolic entities. Global cohesion is sustained and shaped not by international institutions but by a fast-growing electronic commons emerging from grass roots movements around the world. These grassroots movements use the faltering global communications infrastructure to create a new, global commons.

Neither the public sector nor private sector has the resources to mount large-scale projects that might offer a meaningful response to the magnitude of collapse underway. Adaptations are pushed down to the local level of neighborhood and community where people must rely on the people, skills and resources available nearby. Old sources of value—measured in cash, stocks and bonds—have become nearly worthless. New sources of value reside in strong personal connections and access to scarce resources—food, medicine, fuel, etc.—with tangible importance and usefulness.

In the 2040s, much of humanity's story can be told under two headings: the "Great Dying" and the "Great Burning." Although tens of millions of people perished in the previous decade, human die-off escalates as a horrific period of "Great Dying" begins in the 2040s. The carrying capacity of the Earth is estimated to be roughly 2–4 billion people living in middle class, European lifestyles. With a global population approaching 9 billion, we are billions of people over estimated carrying capacity.[110] Humans discover we are no different from the rest of life on Earth facing extinction.[111] Death arrives in many different forms: Waves of disease, famine, and violence over scarce resources.[112]

A tidal wave of death floods the Earth
and stains the soul of the species.

The mathematics of death are unrelenting. With roughly 9 billion people on the planet in the decade of the 2040s and, conservatively, with 10% of the world's population—the poorest of the poor—at greatest risk of dying within this decade, it means that 900,000,000 people could die in this ten-year period. Basic arithmetic translates this figure into an astonishing 90,000,000 people dying *each year*—which is roughly the equivalent of seven holocausts for *each year* of the ten years in this decade.

Waves of death sweep the Earth. The moral and psychological impact of these losses staggers the human psyche and soul as this catastrophe unfolds in real time with high definition media revealing faces and passing lives of countless beings. The immeasurable pain and suffering of the Great Dying tears apart the fabric of human culture and consciousness. The loss, grief and sorrow are incalculable. These wrenching years shred our connections with the past and leave our legacy in tatters.

The magnitude of the tragedy and suffering of the Great Dying transforms the heart and soul of humanity.[113]

The second area of great tragedy and suffering marking this decade is the "Great Burning."[114] Although extreme fires have been burning in localized areas of the world since the 2020s, raging fires throughout the planet have become a dire emergency two decades later. As global warming intensifies areas of drought around the world, the great burning also intensifies.

- Much of the Amazon has dried out and is burning.[115]
- Large swaths of California and the Western United States are chronically on fire, transforming ancient forests into scrubland and brush.[116]
- Large areas in the Los Angeles region burn as do large regions in Texas and Colorado.
- Considerable portions of Mexico are aflame.
- Much of Australia is being incinerated.[117]
- Large regions of Europe—especially southern France, Portugal and the rest of the Mediterranean region—are burning.
- Major portions of India, Pakistan, Iran, and Afghanistan are on fire.
- Regions of northern and southwestern China are regularly in flames.
- Large areas in Africa are chronically ablaze—especially Ethiopia, Uganda, Sudan, and Eritrea.

Instead of labelling our age as the "Anthropocene," in his book *Fire Age,* Professor Stephen Pyne defines it as the "Pyrocene"—a future with upheavals so immense and unimaginable that "the arc of inherited knowledge that joins us to the past has broken . . . a future unlike anything we have known before."[118]

The "Great Burning" and the "Great Dying" symbolize the functional disintegration and disconnection of human civilizations with the past. We are literally no longer able to function as we did before.

Despite the great efforts of previous decades, humanity's evolutionary experiment is failing. The last vestiges of trust in humanity's historical path of material progress are drained from the world.

The powerful elites who dominated the world in prior decades are retreating as the world falls apart around us all. The world eco-crisis is accomplishing what nonviolent action and protest could not—the awakening of humanity's psyche and soul. What is needed above all is a new and purposeful pathway forward. A strong vision and voice for taking us forward is critical.

The people of the Earth are experiencing CTPS—Chronic Traumatic Planetary Stress—an entirely new mindset encompassing the entire human family. The difference between PTSD (post-traumatic stress disorder) and CTPS is that, instead of a relatively brief and confined episode, the trauma is life-long and planetary in scope. There is no escape—the burden of collective trauma permeates the psyche and soul of humanity.

Even while absorbing this decade of immense suffering, people realize our deteriorating biosphere will produce still greater suffering in the coming decades. Many are coping with the social and psychological trauma of being torn from their roots of land, culture, community, and livelihood. Although this has occurred in the past, in the decade of the 2040s, this has become a planetary-scale phenomenon. Consequences of CTPS include:

- Extremely high levels of social anxiety, fear, and protective responses
- Narrowing focus of attention and difficulty concentrating on the big picture
- Emotional numbing and widespread use of alcohol, drugs, and media for escape
- Reactivity, violence, and mood disorders
- Feelings of helplessness, hopelessness, and depression leading to epidemics of suicide

The incalculable suffering of this decade is dissolving old identities and dogmas and leaving many people deeply wounded, both psychologically and socially. Hans Seyle wrote, "Every stress leaves an indelible scar, and the organism pays for its survival after a stressful situation by becoming a little older."[119] At the very time we need to pull together in cooperation as a species, Chronic Traumatic Planetary Stress makes it much more difficult.

The immense suffering of these times is not without merit. In the consumerist pursuit of ongoing happiness, many lost touch with the depths of life—with our souls. Francis Weller has worked with groups, facilitating an authentic encounter with grief, for more than two decades. He writes:

> "To traditional people, soul loss was, without doubt, the most dangerous condition a human being could face. It compromises our vital energy, decreases joy and passion, diminishes our aliveness and our capacity for wonder and awe, saps our voice and courage, and ultimately, erodes our desire to live. We become disenchanted and despondent."[120]

There is a great gift in great grief—this is a passageway to reconnect with our soul. Carl Jung said, "Embrace your grief for there your soul will grow." Unacknowledged sorrows limit contact with the collective soul of our species. As humanity encounters the darkness of our collective losses, we recover contact with our communal soul. Weller further writes that,

> ". . . without familiarity with sorrow, we do not mature as men and women. It is the broken heart, the part that knows sorrow, that is capable of genuine love. . . . Without this awareness . . . we remain caught in the adolescent strategies of avoidance and heroic striving."[121]

Grief challenges the unspoken agreement of consumer society to live "numb and small." Grief is an entrance to the natural, undomesticated aliveness of our soul. Welcoming grief is the secret to being fully alive—the doorway to the wild, untamed vitality of the soul. Naomi Shihab Nye, in her poem "Kindness" writes:

> Before you know kindness as the deepest thing inside,
> You must know sorrow as the other deepest thing.
> You must wake up with sorrow,
> You must speak it till your voice
> Catches the thread of all sorrows,
> And you see the size of the cloth.[122]

The magnitude of the world's fabric of sorrow is immense. We discover what the indigenous soul has always known: We are not separate from the Earth and aliveness is everywhere and in all things. When the Earth is impoverished, we are impoverished in equal proportion.

There is so much for humanity to grieve because the losses are so great: In the Great Dying we have lost millions of precious human beings—sisters and brothers seeking their unique life on the Earth, their potentials unrealized, relationships unfulfilled, talents unexpressed, gifts unreceived by others. We have also lost so much of the rest of life—the plants and animals that bring richness, resilience and beauty into our lives.

With sea level rise, we are losing many of the world's oldest cities established on the seacoasts—Alexandria, Egypt; Shanghai and Hong Kong, China; Jakarta, Indonesia; Mumbai, India; Ho Chi Minh City, Vietnam; Osaka and Tokyo, Japan; London, England; New York and Washington, DC, USA.; and many more.[123]

We are losing lives, cities, cultures, languages, and wisdom. These losses are so foundational to human civilization and our species memory that the grief and sorrow they produce awaken our soul and provide a place of clear knowing from which we can build an authentic future.

The people of the Earth awaken to the understanding of Ubuntu: I am who I am because of who we are. When the "we" is diminished, I am diminished in proportion to the richness of life that has been lost. When we are in contact with our essence, our soul, we are immersed in the larger ecology of aliveness. We share in the kinship of all beings and experience directly the subtle hum and song of all life on the planet.

The sorrows of our wounded world penetrate the small self and awaken our soul. In the grip of overwhelming sorrow for the immensity of our losses, we yearn to return to where we were before grief has overtaken us. Yet, we know we can never go back; instead, we are challenged to accept our fate and discover how this wisdom can transform our pathway into the future.

Collective grief burns through fabrications and facades, and we encounter our raw humanity. In the authenticity of that encounter, we move forward to build new worlds.

In the sorrow of the Great Dying and Great Burning,
we are naked to evolution. Sorrow is no con game.
This is the real world.

When we are claimed by sorrow, we know this world is not a make-believe world. This is the honesty of life itself, and it is to be honored and accepted. Jennifer Welwood is a teacher of spiritual psychology and a poet whose writing speaks to these times:

> My friends, let's grow up.
> Let's stop pretending we don't know the deal here.
> Or if we truly haven't noticed, let's wake up and notice.
> Look: Everything that can be lost, will be lost.
> It's simple—how could we have missed it for so long?
> Let's grieve our losses fully, like ripe human beings,
> But please, let's not be so shocked by them.
> Let's not act so betrayed,
> As though life had broken her secret promise to us.
> Impermanence is life's only promise to us,
> And she keeps it with ruthless impeccability.
> To a child she seems cruel, but she is only wild,
> And her compassion exquisitely precise:
> Brilliantly penetrating, luminous with truth,
> She strips away the unreal to show us the real.
> This is the true ride—let's give ourselves to it!
> Let's stop making deals for a safe passage:
> There isn't one anyway, and the cost is too high.
> We are not children anymore.
> The true human adult gives everything for what cannot be lost.
> Let's dance the wild dance of no hope![124]

Grief takes us beyond hope to the raw truth of reality. In our collective grief, we are called to move beyond our species adolescence, to acknowledge our actual situation, to show up for what is real and to respond as best we can.

The Great Dying cries out for collective maturity
beyond hope or despair—and summons us to step up
and simply take responsibility for doing the work
called for by our times of Great Transition.

Grief reveals the depths—life within life. In the encounter with death, we are ready to turn more fully to life. As we encounter what seems most unbearable, we discover what is most poignantly alive. Grief demolishes pretense and cuts through the superficial happy talk of consumer culture and reveals a stunning aliveness beneath the surface of things. An African proverb says, "When death finds you, make sure it finds you alive."

We have reached a hinge-point in human history where humanity must make choices that will have consequences cascading into the far future. This is evolution in the raw. The Great Dying calls us to a higher level of collective maturity—to reach beyond our species adolescence to take charge of our future.

Humanity wonders whether we "have what it takes"—the maturity to place the well-being of life above our personal interests. Can we engage these difficult times with humility and compassion? Can we talk less and listen more to the suffering of the world? Can we take charge of how we live and work to create a habitable biosphere, understanding that this requires a dramatic change in our manner of living?

Particularly in wealthier nations, a deep psychological crisis has developed as people feel enormous guilt and shame for the devastation of the planet and the diminished opportunities for future generations. Many are in mourning for the Earth and feel that humanity has failed in its grand experiment in evolution. After tens of thousands of years of slow development, many feel that within the span of a single generation we have ruined our chance at evolutionary success and are grieving this lost opportunity. The human community recognizes we face a bleak future of widening ruin and deepening despair unless we collectively rise to this time of challenge.

The suffering, distress, and anguish of these times become a purifying fire that burns through ancient prejudices and hostilities to cleanse the soul of our species. Waves of ecological calamity have reinforced periods of economic crisis, and both have been amplified by massive waves of civil unrest. Momentary reconciliation is followed by disintegration and then new reconciliation. In giving birth to a more conscious, sustainable species-civilization, humanity is moving back and forth through cycles of contraction and relaxation until we utterly exhaust ourselves and burn through the remaining barriers that separate us from our wholeness as a human family.

Finally, we know with unyielding certainty:
we have a choice between extinction and transformation.

In the 2040s, many wondered whether humanity's demise would be a tragedy or a blessing.[125] Are we such a precious contribution to the Earth that we deserve to live and a million other species do not? A deep, moral crisis pervades the Earth. Are we worthy of continued existence? Is there a pathway and purpose that enables us to rise above these tragedies and be worthy of life?

Efforts at reconciliation were begun with a sense of promise and hope, only to fall back in the face of climate chaos and systems breakdown. Is there truly a basis for living together on this small Earth with so many differences? We know the sorrow and divisions of a broken world must be accepted before they can be healed—acknowledging our brokenness is the beginning of the path toward wholeness. Although this is a time of psychological and emotional exhaustion, a critical challenge facing humanity is that of reconciling the many differences between races, genders, incomes, generations, species, religions, and more.

Pushed by dire necessity, innovations in building new kinds of community are growing. People are retrofitting old structures to create new expressions of community ranging from pocket neighborhoods and co-housing to eco-villages of diverse designs. Life-boat communities are proliferating as people recognize that smaller-scale constructions can adapt rapidly to changing circumstances. People recognize the importance of being surrounded with healthy communities, so support for transition towns and sustainable cities is also growing. Simultaneously, tensions are rising as waves of climate refugees seek to move into healthy communities in search of security and survival.

No longer is simplicity of living regarded as a regressive way of life. Low carbon lifestyles and accompanying values are bringing a new regard for community, sufficiency, and kindness. Simplicity of living encourages strong communities of mutual support and survival. As people develop a range of skills that directly contribute to the well-being of their neighbors, they feel their true gifts are being welcomed into everyday life.

Three decades of sustained, local to global communication with Electronic Town Meetings and other forms of civic dialogue have achieved only partial success. As we confront the reality of profound climate catastrophe, we know we cannot retreat; yet, for many, conversations seeking reconciliation and a way forward seem endless and fruitless.

6.4 The 2050s: A Conscious Species–Civilization–The Great Awakening

The Great Dying and the Great Burning leave no doubt that the world of the past is gone, with no hope of returning. Humanity can only move forward after the visceral encounter with deep sorrow that has opened a passageway to our collective soul and a turn toward aliveness. The future is radically unknown and our time of choice is unavoidable. We know intimately the words of poet Wallace Stevens:

> After the final no there comes a yes
> And on that yes the future world depends."[126]

What will be the "yes" of humanity?

At the beginning of the decade of the 2020s, we recognized that building a habitable Earth would require the rapid reduction in CO_2 emissions to net zero emissions by 2050. The fateful decade of the 2050s arrives with the frightening recognition that humanity's efforts, although heroic, were too little and too late. We have not reached this critical goal.[127]

As the unyielding reality of an unfolding climate catastrophe and dark future for humanity hits home, the human community is pushed back upon itself to reconsider how to move forward. Can we transform how we collectively think (our species mind) and how we collectively experience our purpose in living on this Earth (our species soul)?

The last decades have brought shattering climate despair and grief. We have given up the project of trying to reclaim the past.

Can we awaken our species mind and soul to mature ways of imagining and building the future? Do we have the social will and capacity to make this great turning? Joanna Macy sums up the situation clearly:

> "[Are we] . . . serving as deathbed attendants to a dying world or as midwives to the next stage of human evolution. We simply don't know. So, what is it going to be? With nothing to lose, what could hold us back from being the most courageous, the most innovative, the most warm-hearted version of ourselves that we can possibly be?"[128]

Major Driving Trends in the 2050s

- **Global Warming and Climate Disruption:** The goal of zero CO_2 emissions by 2050 is not reached. Global temperatures increase to 3°C (5.4°F) and produce highly disruptive and destructive climate changes.[129] Methane is pouring into the atmosphere, amplifying extreme weather patterns, reducing agricultural productivity, impacting coastal areas with storm surges and hurricanes, and profoundly disrupting the habitats for plants and animals. With unprecedented warming, the oceans are largely depleted of life, the soil is cooked and dry, and ecological breakdowns are widespread as plants and animals cannot adapt to the speed of climate change.

  For several decades we have recognized that, if temperature increases to 3°C, the chance of avoiding four degrees of warming is poor and, if we get to 4°C, it will produce more intense feedback loops and be very difficult to stop temperature rise at 5°C.[130]

  The full climate crisis has arrived.

- **Water Scarcity:** Water stress is expected to impact 52% of the world's population by 2050."[131] With world population approaching 10 billion, this means more than 5 billion people will likely suffer water scarcity.[132] (This estimate ignores the likelihood of a period of Great Dying where a billion or more humans perish.) For many, living has become a miserable struggle for survival in an over-heated and parched world.

- **Food Scarcity:** By 2050, global population is expected to reach upwards of 10 billion, yet food supplies are under enormous stress and are imperiled as the world moves toward an increasingly barren wasteland lacking a rich diversity of plant and animal life. Food demand is 60% higher than it was in 2020, yet, global warming, urbanization, and soil degradation has shrunk the availability of arable land.[133] Recall that each degree Centigrade (1.8° Fahrenheit) of warming is estimated to produce a 10% to 15% decrease in agricultural yields. Therefore, at 3°C, agricultural productivity is falling by 30% to 45% as a result of rising temperatures.

  Compounding the situation, efforts to reduce carbon emissions include reducing the use of petroleum-based fertilizers and pesticides. Unable to prop up agricultural production, food stocks fall further and billions of people are at risk of starvation. "As many as five billion people . . . face hunger and a lack of clean water by 2050 as the warming climate disrupts pollination, freshwater, and coastal habitats. People living in South Asia and Africa will bear the worst of it." [134]

- **Climate Refugees:** Upwards of 300 million climate refugees are expected by mid-century and could be much higher.[135] An enormous number of refugees sets the stage for equally enormous conflicts in the more habitable regions of the planet.

- **World Population:** The world has grown to an estimated 10 billion people by 2057[136] However, this estimate does not factor in the "Great Dying" of the 2040s where roughly 10% of the world's population might perish. The potential magnitude of death in the 2050s seems unimaginable, especially with increasing water scarcity and declining agricultural productivity.

- **Species Extinction:** Habitats for plants and animals around the Earth, on land, in the oceans and in the air are profoundly disrupted at speeds far in excess of their ability to adapt. By mid-century, roughly one-third of all life on the planet is dying with horrendous results. The death of entire species of insects is producing a cascading collapse of the biosphere. The quantity and character of food supplies are dramatically altered.

Grasslands are imperiled. Animals that rely on plants for food are imperiled. The short-term beneficiaries of these die-offs are scavengers on the land—such as cockroaches and vultures—and jellyfish in the oceans.[137]

- **Economic Growth/Breakdown:** By mid-century, the impacts of global warming are dire. Efforts to cut carbon emissions to zero depress economic growth. Faltering growth is compounded by a growing wave of economic breakdowns, bankruptcies and organizational disintegrations. Shortages grow along with hoarding, black-markets, and violence. Traditional sources of value (cash, stocks and bonds) continue to decline while the value of scarce medicine, foods, and fuels rise. Agricultural productivity is continuing to drop as temperatures rise.

  Climate disruptions and massive human migrations are disrupting patterns of trade and production. Consumer society is highly disrupted, fragmented and transitioning to local living economies. The growth mindset of the past is largely replaced by a survival and sustainability mindset with an emphasis on building local resilience in living economies.

- **Economic Inequities:** Extreme inequities persist despite attempts to create a fairness revolution. Although the entire planet is feeling the effects of the whole systems crisis, the impact of global warming is felt most intensely by people who are least responsible for creating it and who are least able to mitigate it. The world's poor live within zones most impacted by global warming and are the most vulnerable to famine, disease, and dislocation.

  Extreme poverty also means people are without essential tools and resources required for contributing, with their own efforts, to the building of an eco-civilization for the Earth. Greater fairness in access to basic technologies and resources is essential to improve the health and productivity of the disadvantaged and to create the foundation for a more sustainable future. Those with access to vital resources recognize that improving the living circumstances of those most impoverished is more than an expression of compassion, it is the way to mobilize a high leverage, grassroots response to climate disruption.

Scenario: How the Decade of the 2050s Unfolds

The Great Dying is continuing as millions of people perish each month. The shadow of needless suffering and killing permeates the world and darkens humanity's outlook on the future. The world is further haunted by the "Great Burning" that is accelerating as runaway global warming picks up speed.

Enormous flows of climate refugees are seeking to move into resource favored areas and this produces strong pushback from local militia. Efforts by local communities to share resources with waves of refugees are soon overwhelmed by the enormous numbers of people and the scale of demands they place on already overstretched systems. Even areas that wanted to be welcoming find themselves challenged far beyond their capacities. Overwhelm produces violent conflicts as people and communities are pressed to the edges of survival. Violence fosters local isolationism and a "build walls" mentality.

Particularly in developed nations, a deep psychological crisis continues to grow as people see the devastation of the planet and diminished opportunities for future generations. Many are in mourning for the Earth and feel that humanity has failed in its grand experiment in evolution. After thousands of years of slow development, many feel that we have ruined our chance for an evolutionary leap forward. The soul of humanity is deeply wounded with moral injury—we have devastated the Earth and violated our intuitive sense of ethics. We face a future of unending bleakness and despair unless we rise collectively to this time of challenge. Do we have the social will and capacity to make this great turning?

The Great Dying that began in the 2040s confronts humanity with the challenge of moving beyond self-centered adolescence and into an early adulthood with care for the well-being of all life. The unyielding reality of death—humans, plants, and animals—pushes humanity to consider the existential significance of life. Why are we here? What is the purpose in being alive?

We are pushed back to our earliest roots and remember indigenous wisdom that describes an infusing life-force throughout the world. We are pushed to consider insights from science that regard the universe as a profoundly unified whole, almost entirely invisible, and permeated by an ecology of consciousness. We see how the views of the world's wisdom traditions are in accord with the views of science—both regard the universe as a regenerative system that is co-arising freshly at every moment. We further recognize how the views of science and wisdom traditions are in accord with the direct experiences of a majority of people who regularly feel our individual lives connect with a larger aliveness.

As we grow in seeing the universe as a unique kind of living system, it transforms our sense of identity and evolutionary journey: Increasingly, we humans regard ourselves as both biological and cosmic beings who are learning to live within and care for the ecology of aliveness. On the one hand, materially privileged societies are moving beyond consumerism as a primary purpose. On the other hand, materially impoverished societies are moving beyond consumerism as a primary aspiration. In breaking the consumer trance of materialism, both are liberated to seek a new world that is sustainable and purposeful.

Lifeways of conscious simplicity are rapidly becoming a core value for living wisely. A deep and conscious simplicity is growing in the world that touches nearly every aspect of life—the food we eat, the work we do, the homes and communities in which we live, and more. We are seeking to heal our relationship with one another, the Earth, and the universe. Conscious simplicity is not simple. In the words of Leonardo da Vinci, "Simplicity is the ultimate sophistication." This lifeway is blossoming with a diverse garden of expressions.

Throughout human history, great differences in wealth and power have been a consistent precursor to civilizational collapse. The preceding decades confirm this pattern. Distraction and delay produced by powerful elites seeking to perpetuate their advantages has resulted in half-hearted measures being taken to remedy great challenges. As a result, "too little and too late"—the lament of nearly all the world's earlier civilizations that have gone extinct—becomes our lament as well. Humanity understands that acceptance of the urgency, depth, and scope of change is the most basic requirement for moving through our time of great transition.

Although the collapse of planetary civilization has advanced significantly, it has not yet extended so completely to the biosphere that we have irreparably damaged the foundations of life from which to rebuild a workable future. We recognize that we are now obliged to reconstruct life from the physical depletion and biological wreckage of a badly injured Earth, but there remain enough remnants from which to rebuild our future, if we step up to the extraordinary level of collective effort and sacrifice now required.

The reconstruction of Earth as a habitable life system will take centuries to accomplish—and we recognize these will be centuries of great learning and deep humility to compensate for the previous arrogance of our adolescent mindset that has so terribly injured the planet.

The question of questions is:
How can the human community rise together
to meet with solidarity the challenges we now face?

We recognize it was communication that enabled humans to evolve from early hunter-gatherers to the verge of planetary civilization, and it will be communication that enables us to become a bonded human community working for the well-being of all. We appreciate that our ability to communicate is the lifeblood of both democracy and planetary civilization. By the 2050s, we are three generations into the communications revolution and, in these decades, we have developed a realistic understanding of how our communication is constantly being manipulated for power and profit. People recognize the internet and television are enabling a new consciousness and empowerment to emerge among the people of Earth—and there is also a deep recognition of the many ways in which these technologies have been manipulated and misused.

Many remember the "Arab Spring" movement that spread across much of the Islamic world in the early 2010s and used the internet to spread messages of reform, which helped to topple the authoritarian government of Egypt and energize activism for democratic reforms throughout the Middle East. However, the same technology was then used by opposing forces to help dismantle the movement.[138]

People also recall that in the US elections of 2016 and 2020, Russia, Iran and other countries sought to interfere in the US elections with fake news and hate speech, flooding the internet and using every major social media platform—Facebook, Twitter, YouTube, and Instagram—to create chaos, conspiracies confusion, and distraction.[139]

The human community has become familiar with countries such as China that strictly control public discourse on the internet by monitoring communications, banning topics from news sites, and limiting personal gatherings that might promote "unhealthy thought." Given decades of experience, by the 2050s there are few illusions that communication is free of intrusions.

The people of Earth have become savvy citizens, skilled in using a wide range of tools to secure the integrity of local to global communication. A widely anticipated "singularity" has emerged in the 2050s. Supercomputers now have such vast computational capacities that they can easily monitor the voting of billions of persons in real time. By combining the power of artificial intelligence with the trusted records of blockchain technologies, systems are available to ensure confidential voting by billions of persons in secure networks.

An "Earth Voice" movement is maturing, bringing confidence that the authentic voice of the people of Earth can be heard and, together, we can choose our pathway ahead. A majority of people are eager for deeper dialogue and feel a growing sense of:

Identity as *Earth citizens*. A larger identity does not diminish other identities of nationality, community, ethnicity, etc. but rather acknowledges the reality of interdependence and responsibility for the well-being of the Earth and our collective future.

Empowerment as *Earth citizens*. Decades of participation in diverse electronic forums have demonstrated that citizen feedback can have a powerful influence on public policy.

Esteem as *Earth citizens*. Despite differences in wealth and privilege, every person's voice and vote count equally in choosing humanity's future.

Solidarity as *Earth citizens*. Decades of trauma and suffering have created new bonds of trust and community. We recognize securing the future is a team effort and that we are all in this together.

Driven forward by grief and pulled ahead by the promise of a healing journey, the global nervous system is awakening a new capacity for collective self-awareness. A new "species mindedness" or Earth-scale reflective consciousness is emerging. We are able to observe ourselves—to know ourselves in the reflective consciousness of our collective mind—and guide ourselves to higher levels of organization, coherence and connection.

Although we have gone to the very edge of ruin as a species, with local to global dialogues we have to pulled back with new levels of collective maturity and insight. After exhausting all hope of partial solutions, we are learning to move in concert toward a future of mutually supportive development. Reaching beneath the chaos and sorrow of these times, we are beginning to discover a deeper sense of community and collective purpose.

In a momentous transition, humanity is awakening to our species-mind and soul. We have been living through the Great Dying and are now maturing into our Great Awakening as an Earth community—a super-organism. We have emptied ourselves of the promises of the consumer trance. Now we are coming together in the stunning aliveness of a new order of humanity. A new human alloy is being forged in the fires of Great Transition.

6.5 The 2060s: Preserving the Future— Choosing Earth

In the 2060s, driving trends are increasing in both the intensity and the pace of change, forcing humanity to either step up and take responsibility for healing a profoundly wounded world or to fully surrender to chaos and collapse:

- **Global warming** has reached a catastrophic level of $3°C$ ($5°+F$). The world's climate is in chaos.

- **Water scarcity** is stressing more than half of the world's population and producing intense conflicts and violence over access to water.

- **Food scarcity** is growing as population increases and productivity falls. Half of the world's population confronts chronic scarcity and famine.

- **Climate refugee** numbers are continuing to grow dramatically. An estimate by Cornell University is that by 2060, an astonishing 1.4 billion people—about one-fifth of the world's population—could become climate change refugees.[140]

- **World population** has grown to roughly 10 billion even with the Great Dying of the previous decade. We are far beyond the regenerative capacity of the Earth. Hundreds of millions of people are still dying from famine and disease.

- **Species extinction** is accelerating as plants and animals are unable to adapt swiftly to the dramatic changes in Earth's climate and weather patterns. The biosphere is collapsing and becoming impoverished.

- **Economic breakdown** is widespread producing hoarding, black markets, and violence. A concern for economic growth has given way to struggles for sheer survival.

- **Economic inequities** persist, thrusting many of the world's former middle class into whirlpools of poverty and calamity.

Scenario: How the Decade of the 2060s Unfolds

The human family recognizes we have come to a choice point in history. The nurturing Earth which supported the rise to a global civilization has been transformed by fires, floods, droughts, famines, diseases and extinctions. Whether the biosphere can be repaired sufficiently to support the rise of a new human civilization is uncertain. We know with certainty that we must work together or effectively perish.

The starting gun of history has truly gone off and we are in a race with a disaster of our own making. One developing catastrophe is in Antarctica, produced by melting glaciers. Thwaites is a glacier the size of the island of Britain, or the state of Florida, and is the most dangerous glacier in the world. Thwaites is melting from underneath and will break away in coming decades. When it slides into the ocean, it will produce a sea level rise of 10 feet or more.[141]

This is enough sea level rise to flood a number of major seacoast cities around the world and, in turn, will amplify mass migrations with a new wave of hundreds of millions of people. By 2060, there is enough momentum in global warming to make the thawing and eventual breaking away of Thwaites irreversible.

We cannot go back and reverse sea level rise; instead, we must go forward with the realization that a new world of flooding seacoasts is now our fate. In the coming century, roughly a third of the major cities along the seacoasts will be submerged and lost to the oceans. Massive urban centers that gave rise to the industrial era are slowly drowning. Many of the economic engines that produced global warming and catastrophe—factories, stores, roads, businesses, and homes—are going underwater.

A slow-moving disaster is playing out over decades. We cannot sprint toward a quick fix; instead, humanity faces a marathon of adaptation that will require sustained effort over generations. An indigenous understanding is being called back into the world. We must look seven generations or more into the future to discover actions in alignment with a sustainable and purposeful future.

Pushed by dire necessity, the world turns toward the large-scale use of climate geo-engineering to limit global warming.[142] Although the undesirable and unexpected consequences of these approaches have limited their use in the past, there is now such urgency that restraints are dropped and two approaches are pursued at a global scale: 1) solar geoengineering techniques to reflect a small proportion of the Sun's energy back into space, suppressing the temperature rise caused by increased levels of greenhouse gases, and 2) carbon geoengineering that employs techniques to remove carbon dioxide or other greenhouse gases from the atmosphere, directly countering the increased greenhouse effect. Because CO_2 capture has not been developed that can work at a planetary scale in the time frame involved, solar geoengineering is the riskiest but most effective in reducing global warming.

With solar geoengineering, a thin shroud of particles fights global warming by imitating the fine ash from volcanic eruptions that deflects solar radiation streaming into the atmosphere. While this shroud of particles suppresses the rapid rise in global temperatures, the reduction of solar radiation creates massive changes in weather systems and rainfall patterns, that are mostly driven by solar energy. With geoengineering, the Asian monsoons, on which 2 billion people depend for their food crops, are shutting down.

Although climate geoengineering is producing major problems for the biosphere, they seem far less dire than the temperature increases they can offset—at least until humanity is able to develop sufficient renewable energy and implement lifestyle changes to live in greater harmony with the Earth. Therefore, despite long-term consequences, geoengineering approaches appear preferable to continuing runaway global warming and are implemented with determination on a planetary scale in an effort to stabilize and then reverse global warming.

A major challenge in mobilizing the human family to support collaboration for a sustainable future stems from the many divides that have endured from the past. Divisions with regard to wealth, gender, religion, race, ethnicity and geography remain; however, the communications revolution has become a powerful force for promoting global reconciliation. Martin Luther King, Jr., said that to realize justice in human affairs, "injustice must be exposed, with all of the tension its exposure creates, to the light of human conscience and the air of national opinion before it can be cured."[143] Injustice and inequities have flourished in the darkness of inattention and ignorance, but with the healing light of public awareness focused on them, it is creating a new consciousness among all involved.

Because everyone knows the whole world is watching as divisions of race, ethnicity, gender, wealth, religion, and more are brought into the court of world public attention, it brings a powerful restorative influence into human relations. With countless resolutions, petitions, declarations, and polls from every region and level of the world, the people of the Earth make their sentiments known—we choose, again and again, to transcend these differences and pull together in cooperation instead of pulling apart in conflict.

A commitment to a sustainable and purposeful future is slowly becoming anchored—visibly, consciously and deeply—in our collective psyche. Pushed by dire necessity and pulled by compelling opportunity, the great turning that humanity has been seeking is gradually emerging from the grief and sorrow of these pivotal decades.

The suffering, distress, and anguish of these times
is a purifying fire that burns through ancient prejudices
and hostilities to cleanse the soul of our species.

Our species mind is, step-by-step, emerging with a recognizable character and temperament. We are progressively awakening and developing a new level of maturity and compassion that rises above the separations of the past.

In stepping back and seeing ourselves as a contentious and yet creative species that has enormous untapped potentials for innovation and kindness, we are giving birth to a functioning species-civilization. An Earth-sized species-organism, comprised of billions of individuals awakening as a collective humanity, moves back and forth through cycles of separation and contraction, surrender and relaxation—until we utterly exhaust ourselves and burn through barriers that divide us from our wholeness. With growing solidarity, we are choosing Earth as our enduring home.

We have burned through much of the isolation of the past to discover a deep, soulful relationship with the Earth, all of her creatures, and other humans. We now live within a growing field of collective awareness that, little by little, has been brought to the surface through decades of suffering and sorrow.

The creative intelligence and immense patience of the living universe is increasingly evident to us. We are crossing a threshold to a new level of collective understanding and consciousness. An immensity of sorrow is cleansing the past. The entire history of the Earth and our species has brought us to this place of opening to a larger identity, a larger humanity, and a larger future. We are in the early stages of seeing ourselves as cells in the body of a super-organism. As the old world has unraveled and fallen apart, a new world is painstakingly being assembled from these fragments.

Millions of humans died in this Great Transition and we make their sacrifice holy—a sacred gift. We vow never to forget their great sacrifice; instead, we make it a sacred gift of gratitude as we learn to live within a larger aliveness. The darkness of death has ignited the flame of soulful aliveness. The sorrows of the past are increasingly seen as purposeful. While still deeply in mourning for the loss of so many lives, so many cultures, so many species and so many irreplaceable resources, in stages we commit to new ways of living that remember and honor all that has been lost, transmuting great suffering into new ways of being together.

Exhausted by the shallow projects of consumerism, we are exhilarated by the deep projects of learning to live in our living universe. We looked into the reality of our extinction and reached for a larger life. We accept our fate—recognizing there is no final truce and lasting harmony and, instead, goodwill and cooperation must be won freshly each day—forever.

Realizing there is no final rest
and that we have the skills and stamina for the
ongoing journey, we rise to a new level of collective
awareness, maturity and responsibility.

With a collective "yes," a meaningful majority of the people of Earth consciously choose preservation of the Earth—and step together onto a new pathway ahead.

Our long-term future is far from secure as we turn to the task of restoring our deeply wounded world and establishing ourselves as a viable species-civilization. A natural capacity for ethical behavior is growing within us. Building on a foundation of conscious reflection and reconciliation, the human community begins to take on the restoration and renewal of the biosphere as a common project, and this promotes a deep sense of global community and connection. A global culture of kindness is emerging.

In choosing Earth, we recall the wisdom of Henry David Thoreau: "What lies before us and what lies behind us are small matters compared to what lies within us. And when you bring what is within out into the world, miracles happen."

6.6 The 2070s and Centuries Beyond: Three Pathways Ahead— An Open Future

At the end of the decade of the 2060s—a half century after hitting an evolutionary wall and running into ourselves as a species with the pivotal questions of who we are and where we are going—these lines of poetry from T.S. Eliot speak volumes:

> We shall not cease from exploration,
> and the end of all our exploring
> will be to arrive where we started
> and know the place for the first time.[144]

After a half-century of turmoil and transition, we "arrive where we started" and see, with unyielding clarity, that we still have three, very different futures before us:

1. Chaos, collapse and a new dark age

2. Authoritarian domination and evolutionary suppression

3. Transformation with a new burst of creative evolution

Although the path ahead remains open, the center of social gravity has shifted decisively in favor of a transformational future leading to a mature, planetary civilization. Yet, we are still learning, maturing, and awakening. The future ahead remains a matter of collective choice—or failure to choose. We either move toward a higher maturity with promise—or we fall back into adolescence and ruin.

In reaching this turning zone, we have not healed the great injury to the Earth. We have not reached a new "golden era" of peace and prosperity. Instead, we are continuing to scramble for survival, coping with the challenges of global warming, the adverse consequences of geoengineering, the immense grief and sorrow of the great dying, the difficulty of settling millions of climate refugees, the enormous task of restoring as many of the plant and animal species that we can; and the colossal undertaking of completing the transition to a renewable energy future.

In choosing Earth as our home, we have taken responsibility for home management and care. Despite this commitment, the future is not settled but remains open to great uncertainty and evolutionary possibility. Which of the three, major pathways will ultimately prevail is still unclear. However, a transformational pathway is providing the primary, orienting direction for the Earth and is supported by an awakening citizenry using the power of collective communication to choose our pathway into the future.

The transformational journey ahead does
not begin after healing the many wounds
humanity and the Earth have suffered.
Instead, it is the long journey of healing that is
the journey home. Coming to our initial maturity
and establishing ourselves as a dynamically stable,
self-referencing, and self-organizing species
is the journey to which we are being called.

Centuries of work lie ahead as we reconcile ourselves with living together and building a thriving future on a severely wounded Earth. We cannot quickly undo the great damage that has been done—it will take centuries of restoration for our Earth to become a welcoming home for the far future. The healing journey ahead is our learning journey. A long process of reconciliation, restoration, and creative adaptation that leads to a viable species-civilization is the pathway home.

Rather than putting these challenging stages behind us,
we are asked to accept and integrate them within us.
This is the source of foundational wisdom
enabling us to endure into the distant future.

After the intense feelings of guilt and self-doubt, we feel a new self-esteem as a species. We feel we have "paid our dues"—the price of admission into the first stage of global maturity—through our immense suffering. Great anxiety as to whether our species would survive is replaced by intense feelings of global community, solidarity, and kinship. "We made it through together," we say. "Our species moved through the time of greatest danger that we could imagine, and we survived. We have truly begun to know ourselves as a human family with all of our faults and idiosyncrasies. We know there is to be no rest—that forever we must work to reconcile ourselves with ourselves—and we also know that we are equal to the challenge."

"Looking multiple generations into the future, we know we must follow through persistently, inclusively, and tirelessly. With kindness, maturity and patience, we are willing to return again and again to the commitments already made."

"We have not established ourselves in a miraculous, new golden age. What we have accomplished, however, is momentous: We have reached a place of mature, collective understanding as a diverse and still contentious species. We know that we must work together, forever, if we are to not perish from this Earth but, instead, find our way in learning to live in the living ecology of the Earth and universe."

We begin the next stages of our long journey by recognizing there is as much to learn on the path of coming into meaningful connection as there has been on our journey of separation. It took more than 10,000 years for humanity to learn and grow from the beginnings of the agricultural revolution to our time of great transition. In turn, we expect that it will take many centuries for us to establish ourselves as an enduring species civilization.

Beyond our time of great transition are three additional stages lie ahead as we seek to establish ourselves as a mature species-civilization that can continue into the deep future. The stages of reconciliation, restoration, and creative adaptation are summarized below. These stages are explored in much greater length in my book *Awakening Earth* that is available as a free download on my website: www.DuaneElgin.com

Establishing Stage I: Reconciliation

Although all of the initiatives essential for moving through the time of great transition are vital, perhaps the most foundational requirement is for reconciliation. Reconciliation movements that began during transition must now be fulfilled if there is to be a secure foundation for the future. Deep, ongoing work—lasting many decades or even centuries—is necessary to heal the wounds and divisions of race, religion, gender, wealth, and more.

Reconciliation continues to deepen with truth telling and public recognition of the many divisions that separate us from our wholeness. Beyond public recognition and acceptance of responsibility for the wrongs of the past is the need for sincere, public apologies. Forgiveness does not mean forgetting but collective learning. Public expression of apology and responsibility are then followed by authentic restoration—genuine efforts to make right the wrongs of the past. With reconciliation, we begin to build a future of mutually supportive development.

Where the urban-industrial era mindset with its "thinking consciousness" brought an assertive, materialistic and self-promoting orientation, the "reflective consciousness" awakened during the process of transition brings forth a more receptive, trans-material, and relational orientation—a shift from masculine to feminine qualities. Embracing the feminine archetype—which tends to be more open, allowing, forgiving, and integrating—supports people in moving beyond the separative and competitive mind-set of the previous era and advancing a caring and cooperative consciousness that is necessary for building a sustainable and purposeful future.

Establishing Stage II: Restoration

Although restoration initiatives have been underway since the 2020s, the Earth has been so badly wounded over these decades that there is great need for deep restoration if humanity is to have a resilient, ecological foundation from which to rise to our higher potentials as a species-civilization. The preceding stage of reconciliation makes it possible for people to give full effort to rebuilding and restoring our wounded world. Sustained efforts to repair the ecology of the Earth establish the foundation for a highly functioning species-civilization.

Human culture and consciousness are growing into a cohesive community of life. People feel a new depth of connection with nature and sacred regard for all life. Cross-cultural learning and planetary celebrations flourish in this era of mutually supportive development. We know intuitively that we are not separate from the larger ecology of life. As we participate in the seamless fabric of creation, it naturally awakens compassion for the rest of life and a natural sense of ethics grows within us. With a maturing consciousness, we recognize that we all share in whatever measure of happiness or sorrow is being created.

Establishing Stage III: Creative Adaptation

With the restoration of the biosphere well underway, a new level of human creativity is liberated. Civilizing efforts move beyond maintaining ourselves to surpassing ourselves as we see evolution as an artistic and spiritual adventure as much as a material enterprise. The culture and consciousness of a maturing planetary civilization continuously balances creative diversity with sustainable unity.

Given the creative tumult, the bonding achieved in the previous stages is essential to keep the world from swirling out of control and tearing itself apart. Humanity is learning to balance the drive for creative diversity with the need for sustainable unity. A dynamically stable and self-organizing species-civilization is blossoming—a magnificent flowering of the living universe.

• • •

The dynamic integration of the preceding three stages represents both the completion of a long process of development and the foundation for a new beginning. Humanity has acquired a sufficient measure of reconciliation, restoration and creative adaptation to establish ourselves as a mature planetary civilization that can endure into the deep future.

Lessons from the Great Transition Scenario

The scenario of Great Transition represents an exercise in social imagination. One paramount conclusion emerges from this exploration:

Once we see what lies ahead in our social imagination,
we don't have to enact this future in physical reality.
By mobilizing our capacity for collective imagination,
we can consciously envision where we are headed
and then choose a more promising pathway forward.

There is no need to wait decades before taking decisive action. If we can visualize the imperative for action created by terrible suffering just a few decades into the future, we do not have to postpone our efforts and enact that suffering in actual experience—we can, instead, digest that reality and act creatively NOW.

Mobilizing our capacity for social imagination, we see that we are depleting and impoverishing the precious ecology of life. Recognizing this, we also recognize we have the choice to preserve what remains and to restore the biosphere, NOW.

We are in the midst of a global communications revolution and have the tools to collectively visualize and choose a new pathway ahead. By mobilizing our capacity for social imagination and collective communication, we have the choice to grow up and move into our species maturity NOW.

We can visualize and communicate a new journey ahead in electronic conversations, films, rituals, art, stories, books, music, dance, poetry and more. We can portray, in detail, the changes and actions that lead to a more mature and sustainable future. By visualizing the practical changes required for a great transition, we can swiftly move to realize that pathway ahead. That is the focus of Part IV that follows.

PART IV:

Foundations of Great Transition

7. Turning Toward Aliveness

Humanity's great transition and turn toward aliveness will not happen automatically. This is a demanding process for us individually and collectively. It is primarily through grieving for the great losses that are now occurring, as well as those that lie ahead, that we will awaken and turn toward the promise of aliveness. If we simply drift along, a catastrophic future will surely unfold. Continuing business as usual is likely to produce either a future of authoritarianism or a descent into a new dark age for humanity.

A great transition will require an extraordinary level of collective effort, conscious choice, and creative cooperation among the human community. Nothing short will suffice. The transition ahead is so astonishingly difficult that we may seem destined for defeat. However, we were created for these times and have the capacity to make conscious choices to realize a thriving and surpassing future. Simone de Beauvoir wrote, "Life is occupied in both perpetuating itself and surpassing itself; if all it does is maintain itself, then living is only not dying."[145] Our great turning is a time to choose, together, a future beyond "only not dying" and to surpass ourselves by becoming more fully alive.

Albert Einstein famously said that we cannot solve a problem with the same mindset that created the problem; rather, we have to look freshly from a larger mindset to see the solution.

Our world crisis asks us to
look freshly at our mindset and ask:
"Is there an experience of life that is so widely shared
that it can draw us together in a shared journey
into a thriving future?"

The answer is a straightforward "Yes." Beneath our many differences is the common experience of simply being alive, and this mysterious reality provides an unshakable foundation for humanity to come together in a common journey of transition and transformation.

One of the world's foremost scholars of humanity's wisdom was Joseph Campbell. I had the privilege of co-authoring a book,

Changing Images of Man, with him in 1974. Our task was to explore the deep archetypes drawing us into the future in these transitional times.[146] In a revealing interview, Campbell was asked if the deepest quest of humans is the "search for meaning." He replied:

> "People say that what we're all seeking is a meaning for life. I don't think that's what we're really seeking. I think that what we're seeking is an experience of being alive, so that our life experiences on the purely physical plane will have resonances with our own innermost being and reality, so that we actually feel the rapture of being alive."[147]

Campbell further said that, "I think what we are looking for is a way of experiencing the world that will open to us the transcendent that informs it, and at the same time forms ourselves within it." We are seeking a direct encounter with life itself as a way of connecting with our own aliveness. Eckhart Tolle writes: "There are three words that convey the secret of the art of living, the secret of all success and happiness: *One With Life*. Being one with life is being one with Now. You then realize that you don't live your life, but life lives you. Life is the dancer, and you are the dance."

Psychologist and philosopher, Erich Fromm, has written of the joyful reciprocity in sharing our experience of aliveness.[148] He says that our experience of aliveness is the most precious gift we can share with others. We offer the most authentic expressions of what is alive within ourselves—our direct experience of life in all of its richness—our gratitude and fears, understanding and curiosity, humor and sorrow. In sharing our aliveness, we enrich the life of others. We awaken their sense of aliveness by sharing our own experience of being alive in the moment. We do not share with the intention of receiving something from others; instead, the sharing is, in itself, a gift of ourselves that awakens a reciprocal aliveness in others that returns to us in a mutually enhancing flow.

This is not abstract philosophy but the visceral experience of simply and directly being alive to our unique sense of ourselves. At 90 years of age, the words of Florida Scott-Maxwell describe this view powerfully: "You need only claim the events of your life to make yourself yours. When you possess all you have been and done, you are fierce with reality."[149]

*As we awaken to the aliveness at the core of our being,
we are simultaneously connecting with the
aliveness of the universe.*

Aliveness costs nothing and is freely given as our birthright. The experience of aliveness is here and available to us at all times. Aliveness is an embodied, powerful, and universally shared experience. To illustrate, I asked participants in a learning community I co-facilitate to describe what "being fully alive" means to them. Responses were immediate and direct: "Being in the flow; Mind coming home to body; Feeling the full range of emotions; Living on purpose and without expectation; Playful and joyful communion; Giving full expression of soulful gifts; Full engagement with life, Deep connection with nature;" and more. [150]

A life path of developing our full aliveness may be dismissed as wishful fantasy by those living within the mindset of materialism and consumerism—a mindset that views the universe as consisting primarily of dead matter and empty space. This view is changing. The mindset of materialism is being transformed by new findings from science, enduring insights from wisdom traditions, and the direct experience of a large portion of humanity. By integrating diverse sources of understanding, we can reach beyond the shallow assumption of materialism.

*Aliveness is the new—and ageless—experience
that offers humanity a place of common meeting,
mutual sympathy, and collective healing.*

We are discovering the universe is an astonishingly alive, unified, "super-organism" permeated with a knowing presence or consciousness. This is such a foundational shift and is so important to humanity's future that it merits exploration here (many of the following themes are explored more fully in my book, *The Living Universe*).

In recent centuries, the task of creating urban-industrial societies has engaged a large portion of humanity in becoming ever more differentiated and empowered as individuals. This has taken many persons—although certainly not all—away from direct connection with the experience of aliveness.

Throughout history, a deep thread of understanding has endured that regards the universe as alive and ourselves intimately connected within it. The story of separate beings who are living in a dead universe without deeper meaning or purpose is now ending. We are growing into a relational world infused with aliveness—a foundational theme of this book.

To appreciate this perspective, it is important to recognize the thread of aliveness in human experience that has run throughout history and can be found in the wisdom of Indigenous cultures, spiritual traditions, deep ecology, people's direct experience and much more.

Indigenous Wisdom

Our closest connection with the earliest understandings of ancient peoples comes from Indigenous traditions with deep roots that extend far into humanity's past. Indigenous wisdom is extremely important for connecting modern humans with the wisdom of our ancestors, wisdom that enabled them to endure incredibly harsh conditions for several hundred thousand years. How do people who continue to uphold these ancient traditions experience life and the world? Here are illustrative quotes from a range of Indigenous cultures that reveal a subtle and refined understanding of a world of aliveness:

The Ohlone Tribe lived from the San Francisco Bay Area to Monterey.

"For the Ohlone, 'religion' was pervasive, 'like the air.' Nature was seen to be alive and shimmering with energy. Because everything was filled with life, power was everywhere and in everything. Every act was a spiritual act because it engaged the worlds of power. All tasks . . . were done with a feeling for the surrounding world of life and power."[151]

The Koyukon Tribe of northern central Alaska

"The Koyukon live 'in a world that watches, in a forest of eyes.' They believe wherever we are, we are never truly alone because the surroundings, no matter how remote, are aware of our presence and must be treated with respect."[152]

Sarayaku Kichwa, of the Ecuadorean Amazon jungle

"Believe that 'Everything in the jungle is alive and has a spirit.'"

Luther Standing Bear, Lakota Sioux from the region of North and South Dakota

"There was no such thing as emptiness in the world. Even in the sky there were no vacant places. Everywhere there was life, visible and invisible, and every object gave us great interest in life. The world teemed with life and wisdom; there was no complete solitude for the Lakota."[153]

A common intuition and insight are found in Indigenous sources around the world—our life exists within a larger aliveness. A living presence permeates the world and naturally includes what is now called consciousness and described, for example, in Indigenous wisdom as a "forest of eyes" that is aware of our presence, no matter who or where we are. A related intuition is that a life-force or "sacred wind" blows through the universe and brings with it a capacity for awareness and communion with all of life.

The scope and depth of Indigenous wisdom can be appreciated from the perspective of "three miracles" that have been attributed to both Native American cosmology and Western philosophers. The first miracle is that anything exists at all. The second miracle is that living things exist—plants and animals. And the third miracle is that living things exist that know they exist—and that is primarily ourselves as humans. People in wealthier societies tend to focus attention on the third miracle of ourselves and the miracle that we can be conscious of our existence. What is often overlooked in "advanced" societies is the first miracle—that anything exists at all. Yet, it is the first miracle that embraces the totality of existence and calls us to awaken to a vastly larger scope of mystery, appreciation and regard than the third miracle considered alone.

The Jungian philosopher Anne Baring describes how it is difficult for materialist cultures, that assume the universe is non-living or dead at the foundations, to enter into the experience of Indigenous cultures and their fundamental understanding that "the life of the Cosmos, the life of the Earth and the life of humanity were one life, permeated and informed by animating spirit."[154] She writes that the great revelation of our time is that "we are moving from the story of a dead, insentient cosmos to a new story of a Cosmos that is vibrantly alive and the primary ground of our own consciousness."[155]

Spiritual Wisdom

Consistent with Indigenous views, an astonishing insight regarding the nature of the universe is found in diverse spiritual traditions and is only now being recognized. Most spiritual traditions share the understanding that the universe is infused with a living presence and is continuously arising anew at every moment. A broad spectrum of spiritual traditions shares this insight that the universe is emerging continuously as an undivided whole in an unutterably vast process of awesome precision and power:

Christianity: *"God is creating the entire Universe, fully and totally, in this present now. Everything God created . . . God creates now all at once."*[156]
—Meister Eckhart, Christian mystic

Islam (Sufi): *"You have a death and a return in every moment . . . Every moment the world is renewed but we, in seeing its continuity of appearance, are unaware of its being renewed."*[157]
—Jalāl ad-Dīn Muhammad Rūmī, 13th century Sufi teacher and poet

Buddhism (Zen): *"My solemn proclamation is that a new Universe is created every moment."*[158]
—D.T. Suzuki, Zen teacher and scholar

Hinduism: *"The entire Universe contributes incessantly to your existence. Hence the entire Universe is your body."*[159]
—Sri Nisargadatta, Hindu teacher

Taoism: *"The Tao is the sustaining Life-force and the mother of all things; from it, all things rise and fall without cease."*[160]
—Lao Tsu, founder of Taoism

Beneath differences of language and history, a common understanding is revealed—the Universe is a living system emerging continuously as a fresh creation at every moment. We are an inseparable part of this regenerative process. We are paradoxical beings who are both entirely unique and profoundly connected within the regenerative Universe.

Nature Wisdom

Another source of wisdom found throughout history is nature's living presence. The natural world blossoms with a mysterious aliveness that can be summarized as follows:

- There is a creative, living presence beneath, within, and beyond all material things
- Our invisible essence ("soul") has a natural capacity to connect with this living presence
- A core project for humanity is to discover our union with the creative aliveness of the natural world.

From the wisdom of mystics, poets and naturalists there is a common recognition of the infusing aliveness of nature's wisdom.[161]

Heaven is under our feet as well as over our heads.
    —Henry David Thoreau[162]

The deeper we look into nature, the more we recognize that it is full of life. . . . From this knowledge comes our spiritual relationship with the universe.
    —Albert Schweitzer[163]

This land is the house we have always lived in.
    —Linda Hogan[164]

Going to the mountains is going home.
    —John Muir[165]

And into the forest I go to lose my mind and find my soul.
    —John Muir[166]

Nature is not a place to visit. It is home.
    —Gary Snyder[167]

Not just beautiful, though—the stars are like the trees in the forest, alive and breathing. And they're watching me.
    —Haruki Murakami[168]

I believe a leaf of grass is no less than the journey-work of the stars.
    —Walt Whitman[169]

The goal of life is to make your heartbeat match the beat of the universe, to match your nature with Nature.
　　—Joseph Campbell[170]

If you wish to know the divine, feel the wind on your face and the warm sun on your hand.
　　—Buddha[171]

One touch of nature makes the whole world kin.
　　—John Muir[172]

I believe in God, only I spell it Nature.
　　—Frank Lloyd Wright[173]

Nature is the most clear and contactable way to feel Life. The Vibration of Nature is so very tangible and so when I look at the clouds, smell the fresh air, watch the insects, listen to the birds and salute the trees. I feel part of it All. And we All are part of it . . . ALL!
　　—Mira Michelle[174]

We are living expressions of a living universe that is growing and discovering itself through its own creations. George Washington Carver, a great educator and botanical researcher, said that "If you love it enough, anything will talk with you."[175] When we experience our love for the life that surrounds us, we feel at home in the universe—we feel that we belong here, and recognize that we are on a purposeful journey of learning to live in the living Universe. The experience of oneness with all of life is transformative and brings with it a deep sense of satisfaction and trust in life.[176]

It is important to honor the role of the deep feminine in this ancient wisdom. Masculine, patriarchal culture has lost touch with much of the aliveness, interconnectedness, and generative power of nature. To awaken to the living Universe is to open to the feminine aspect of our being and a relational approach to living. Theologian Sallie McFague has written, "The universe is the body of God."[177] Cynthia Bourgeault, an Episcopal priest, writes, "We get to participate in [cosmic fullness] freely, fully, here and now, simply because each one of us is a tiny shareholder in the divine aliveness."[178]

Wisdom of Direct Experience

How widespread is the experience of a permeating aliveness and deep unity in everyday life? How often do people feel aliveness and intimate connection with nature and the larger world? Here are scientific surveys that explore this pivotal question:

- A global survey involving 7,000 youths in 17 countries was taken in 2008. It found that 75% believe in a "higher power," and a majority say they have had a transcendent experience, believe in life after death, and think it is "probably true" that all living things are connected.[179]

- In 1962, a Gallup survey of the adult population in the US found that 22% reported having awakening experiences revealing our intimate connection with the Universe. By 1976, Gallup reported this had grown to 31%. By 1994, a *Newsweek* survey found this had grown to 33%. By 2009, the percentage of the population reporting a "moment of sudden religious insight or awakening" had grown dramatically to 49% of the adult population.

Figure 6: Growth in Awakening Experiences in US: 1962-2009 by Percentage of Population[180]

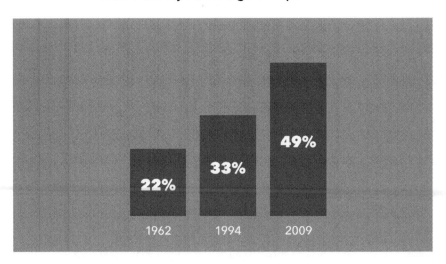

- In a national survey of the US in 2014, nearly 60% of adults reported they regularly feel a deep sense of "spiritual peace and wellbeing," and 46% say they experience a deep sense of "wonder about the Universe" at least once a week.[181]

- In concert with these experiential changes, there has been a dramatic increase in the use of meditation in recent years. From a new age novelty in the 1960s, this has grown into a mainstream movement in the 21st century. The percentage of adults who meditate is growing rapidly—from an estimated 4% of the US population in 2012 to more than 14% just five years later (2017).[182] Now, meditation along with diet and exercise are considered mainstream activities for health and well-being.

These surveys show that awakening experiences of communion and connection with the aliveness of the Universe are not a fringe phenomenon but, instead, are familiar encounters for a large portion of the public. Humanity is measurably waking up to a new (and ageless) view of ourselves as inseparable from the larger Universe.[183]

Awakening to an intimate connection with the unity and intelligent aliveness of the Universe is often accompanied by feelings of joy, boundless love, and the presence of a subtle, radiant light. To illustrate, below is a classic account of a spontaneous awakening experience. While an undergraduate student, F.C. Happold had this experience of communion with the permeating aliveness of the Universe:

"There was just the room, with its shabby furniture and the fire burning in the grate and the red-shaded lamp on the table. But the room was filled by a Presence, which in a strange way was both about me and within me, like light or warmth. I was overwhelmingly possessed by Someone who was not myself, and yet I felt I was more myself than I had ever been before. I was filled with an intense happiness, and almost unbearable joy, such as I had never known before and have never known since. And over all was a deep sense of peace and security and certainty."[184]

Turning from spontaneous awakenings to the intentional exploration of consciousness, for thousands of years, pioneering individuals have been investing their lives in solitude and sustained meditation to investigate directly the nature of reality. What these explorers of consciousness have discovered is not a grey, machine-like hum of a non-living Universe but, instead, an ocean of unbounded love, light, and creative intelligence whose nature is beyond the reach of words.

When our personal aliveness becomes transparent to the aliveness of the living Universe, experiences of wonder and awe emerge naturally. As we open into the cosmic dimensions of our being, we feel more at home, less self-absorbed, more empathy for others, and an increased desire to be of service to life. These shifts in perspective are immensely valuable for building a sustainable and purposeful future.

Scientific Understanding

Until recently, any suggestion that the universe could be a unified, living system was regarded as fantasy by mainstream science. Now, with the findings of quantum physics and more, the ancient intuition of a living Universe is being reconsidered freshly as science cuts away superstition to reveal the cosmos as a place of unexpected wonder, depth, dynamism and wholeness as a living system.[185]

- **A Unified Whole:** In the last several decades, scientific experiments have repeatedly confirmed the Universe is a deeply unified system that is able to communicate with itself instantly across distances that seem impossibly vast. To illustrate, at the speed of light, it takes more than eight minutes for a photon to travel from the sun to the Earth and more than 14 billion years to travel across our visible Universe. Yet, quantum physics demonstrates these vast distances are traversed and connected instantaneously. Additionally, science no longer views the Universe as a disconnected collection of planets, stars, and other matter; instead, the Universe is fully unified and connected with itself at every moment.

- **Mostly Invisible:** In a stunning challenge to the paradigm of materialism as being all there is, scientists now think the overwhelming majority of the universe is invisible and not material! Scientists estimate that approximately 96% of the known universe is invisible to our physical senses with 73% comprised of "dark" (or invisible) energy and 23% comprised of "dark" (invisible) matter.[186] The biological aspect of ourselves is a manifestation of the 4% of the universe that is visible matter.

This new understanding from science connects with humanity's original perception that, underlying the physical world, there is a vastly larger invisible world. We live within a vast ocean of unseen energy of incomprehensibly immense power.

Figure 7: Composition of Universe by Percentage:
Invisible Energy and Matter Contrasted with Visible Matter[187]

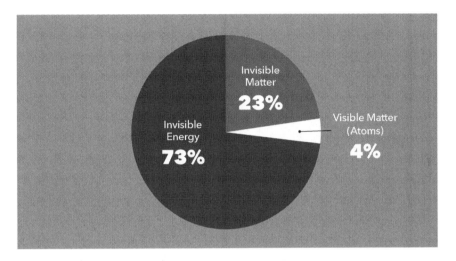

- **Co-Arising Universe:** At every moment, the entire Universe is emerging freshly as a singular orchestration of cosmic expression. Nothing endures. All is flow. In the words of the cosmologist Brian Swimme, "The Universe emerges out of an all-nourishing abyss not only fourteen billion years ago but in every moment."[188] Despite outward appearances of solidity and stability, the Universe is a completely dynamic system.

- **Consciousness at Every Scale:** An ecology of consciousness is increasingly thought to permeate the Universe. Scientists are finding evidence for consciousness or a knowing capacity or centering subjectivity throughout the Universe. The physicist and cosmologist Freeman Dyson has written that, at the atomic level, "It appears that mind, as manifested by the capacity to make choices, is to some extent inherent in every electron."[189] Max Planck, developer of quantum theory, stated, "I regard consciousness as fundamental. I regard matter as derivative from consciousness. We cannot get behind consciousness. Everything that we talk about, everything that we regard as existing, postulates consciousness."[190]

- **Freedom at the Foundations:** Quantum physics describes reality in terms of probabilities, not certainties. Uncertainty and freedom are built into the very foundations of existence. We have great freedom to act within the limits established by the larger web of life.

- **Able to Reproduce Itself:** A vital capacity for any living system is the ability to reproduce itself. A growing view in cosmology is that our Universe reproduces itself through the functioning of black holes. Physicist John Gribbin writes, "Instead of a black hole representing a one-way journey to nowhere, many researchers now believe that it is a one-way journey to somewhere—to a new expanding Universe in its own set of dimensions."[191]

A new picture of our Universe is coming into focus. Modern science and ancient wisdom converge to affirm a stunning observation made by Plato more than two thousand years ago: "The universe is a single living creature that contains all living creatures within it." Life within life! Our aliveness is inseparable from the larger aliveness of a living cosmos. The Universe is a living, cosmic hologram—a unified, "super-organism" that is almost entirely non-physical in nature, that is continuously regenerated at each moment, and that includes consciousness, or a knowing capacity, as a fundamental aspect. We live within a living Universe permeated with a creative intelligence that unfolds evolution from within itself. This view is congruent with that of Albert Einstein who frequently referred to his belief system as "cosmic religion."[192]

There is a helpful precedent for viewing the Universe as a unified, living system. The idea that the Earth is a single, living entity—Gaia—was initially ridiculed and rejected but, increasingly, is seen as a discerning and realistic understanding. In a similar way, the idea that the Universe is a single, living system—Meta-Gaia—is a view dismissed by shallow and separative materialism but, increasingly, is validated by findings from the frontiers of science.

Aliveness is both foundational and emergent.
Aliveness is at the foundation of all existence
and the highest expression of all existence.

Aliveness is both the underpinning of existence and the summit of existence. Yet, in the modern era of materialism, we tend to regard our aliveness as separate from the aliveness at the ground of all existence. A famous quote from Albert Einstein challenges the view of separation:

> "A human being is part of the whole called by us 'Universe,' a part limited in time and space. We experience ourselves, our thoughts and feelings as something separate from the rest. A kind of optical delusion of consciousness. The quest for liberation from this bondage is the only object of true religion."[193]

Thomas Berry, scholar of the world's religions, describes the inseparable connection of the individual and the universe: "We bear the universe in our being as the universe bears us in its being. The two have a total presence to each other and to that deeper mystery out of which both the universe and ourselves have emerged."[194] How extraordinary: An invisible aliveness is creating and sustaining our universe, patiently holding it within its spacious embrace for billions of years while growing ever more conscious expressions of aliveness that are increasingly able to look back with a reflective consciousness and appreciate their origins.

As we learn to recognize our experience of aliveness, and as we encounter the aliveness at the foundation of the Universe as felt experience—as life meets life—a window opens and there naturally arises awakening experiences. When our experience of aliveness connects with the greater aliveness at the ground of all that exists, we recognize, as direct experience, that we are part of the great wholeness of life. This is who we are: both a unique expression of aliveness and an inseparable part of the greater aliveness. We are both—a unique biological aliveness and an inseparable part of the cosmic aliveness. In Thomas Berry's description, the two have a total presence to each other and to the deeper mystery out of which both arise. We are both biological and cosmic in nature—we are "bio-cosmic" beings.

Aliveness and Humanity's Future

When we bring these threads of wisdom together—Indigenous understandings, spiritual traditions, nature wisdom, direct experience, and scientific evidence—how do they inform humanity's future?

Aliveness is our most universal experience and transforms how we relate to our evolutionary journey. If we regard the Universe as no more than physical matter that is dead at its foundation, then feelings of existential alienation, anxiety, dread, and fear are understandable. Why seek communion with the cold indifference of lifeless matter and empty space? If we relax into a dead Universe, we will simply sink into existential despair, so better to live on the surface of life.

However, if we live in a Universe that is almost entirely invisible and brimming with aliveness, then feelings of subtle connection, curiosity, and gratitude are natural. When we see ourselves as participants in a cosmic garden of life that has been developing patiently over billions of years, our regard for the Universe shifts from indifference, fear and cynicism to curiosity, love, and awe. Humanity's future pivots on which understanding prevails and the choices that naturally follow.

Being unflinchingly realistic, it does not seem likely we will turn away from a path of separation—with its growing inequities, overconsumption of resources, and deep injury to the Earth—unless we discover a pathway into the future that is so truly remarkable, transformative, and welcoming that we are drawn together by the felt presence of its invitation. This pathway must be so compelling as a felt possibility that we are drawn into exploration in the present moment. That pathway is being revealed by insights converging from science and the world's wisdom traditions.

We are discovering that, instead of struggling
for meaning and a miracle of survival in a dead Universe,
we are being invited to learn and grow forever
in the deep ecologies of a living Universe.

To step into the invitation of learning to live in a living Universe represents a journey so extraordinary, it calls us to heal the wounds of history and to realize a remarkable future we can only attain together. As we open into the cosmic dimensions of our being, we feel more at home, less self-absorbed, more empathetic toward others, and increasingly drawn to be of service to life. These shifts in perspective are immensely valuable for building a sustainable future.

To step into the invitation of
learning to live consciously in a living Universe
is to begin a new chapter in humanity's evolution
that transforms our sense of identity,
evolutionary purpose, and ethics for living.

Where the mindset of the techno-industrial era washed the life out of nature and left a machine-like cosmos filled mostly with dead matter and empty space, in the emerging era, people are awakening to the intuition that a living presence permeates the universe. In the spacious mirror of a reflective consciousness we begin to catch glimpses of the unity of the interwoven fabric of the cosmos and our intimate participation within the living web of existence. No longer is reality broken into relativistic islands or pieces. If only for brief moments at a time, existence is glimpsed and known as a seamless totality.

Touching the aliveness of the universe for even a few moments can transform our lives. The renowned Sufi poet, Kabir, wrote that he saw the universe as a living and growing body "for fifteen seconds, and it made him a servant for life."[195]

No matter how mundane the circumstance, no matter how seemingly trivial the situation, we can always become aware of the subtle aliveness and consciousness within and around us. We can glimpse the living universe in small ways—perhaps in the golden light of a late afternoon or in the luster of an old wooden table that shines with an inexplicable depth and glow. We can also witness the buzzing aliveness of existence in places that may seem far removed from nature—a room filled only with plastic, chrome steel, and glass will fiercely display aliveness in the raw.

In the gentle contemplation of any part of ordinary reality we can catch glimpses of the great hurricane of energy that blows with silent force through all things and, with a "forest of eyes," is aware of our existence. Empty space will also disclose that it is an ocean of dancing aliveness—a subtle symphony of transparent architecture actively providing a context for matter to present itself.

Bio-Cosmic Identity

From a living Universe perspective, our identity is immeasurably larger than our biological selves. Recall that 96% of the known universe is comprised of invisible energy. Because we are an inseparable part of the universe, it suggests that 96% of who we are consists of invisible energies. We are vastly more than skin-encapsulated, material beings. Our biological existence is inseparable from the larger, non-material aliveness of the living Universe.

Aliveness is not far away in some distant galaxy. It is right here, present everywhere—so close that it is between our fingers and even between our thoughts. Aliveness fills our physical body and is who we are. Seeing ourselves as part of the invisible fabric of creation awakens our sense of connection with, and compassion for, the totality of life. Cosmologist Brian Swimme explains that the intimate sense of self-awareness we experience bubbling up at each moment "is rooted in the originating activity of the Universe. We are all of us arising together at the center of the cosmos."[196]

From the perspective of materialism and its view of a non-living universe, we are no bigger than our physical bodies. Now we are discovering that we are deeply connected participants in a living, regenerative Universe. Spiritual teacher, Rupert Spira, writes: "Your body is a flow, not an object. Don't allow a solid, separate body to crystalize out of this fluid ocean of consciousness. All suffering is a result of believing you are a solid, temporary entity, with the same destination as the body."[197] Another teacher Eckhart Tolle writes, "You are not IN the universe, you ARE the universe—an intrinsic part of it. Ultimately you are not a person but a focal point where the universe is becoming conscious of itself. What an amazing miracle."[198]

Awakening to our larger identity as both unique and inseparably connected with a co-arising Universe transforms feelings of existential separation into experiences of subtle communion as bio-cosmic beings. We are far richer, deeper, more complex, and more alive than we ever thought. To discover this in our direct experience is to enter a new age of exploration and discovery.

Coming to this felt understanding is an immense challenge. The highly separate sense of self—rejects the understanding that we arise from the wholeness of a regenerative Universe. Moving beyond delusions of separation requires a relinquishment of our limited sense of self that, until experienced, appears suicidally threatening. Until we surrender the small "self" to this invitation, we cannot experience the unbounded "Self."

Of course, once we experience our boundless nature, there is no going back to complete separation. The crux of the challenge is for enough members of the human family to overcome their fear of this new (and ageless) understanding of reality and to begin their journey home.

Cosmic Purpose

To be born as a human being is a rare and precious gift. A quote attributed to Blaise Pascal speaks clearly: "The goal of life is not happiness, peace, or fulfillment, but *aliveness*."[199] Howard Thurman, author, philosopher, theologian, and civil rights leader famously said, "Don't ask what the world needs. Ask what makes you come alive, and go do it. Because what the world needs is people who have come alive."[200] We discover our aliveness by fully inhabiting our seemingly "ordinary" lives. While we have the gift of a body to anchor our experience, it is important to recognize our bio-cosmic nature.

We are bio-cosmic beings:
Our bodies are biodegradable vehicles
for acquiring soul-growing experiences.

As compostable conduits for learning experience, our bodies are the current expressions of a creative aliveness that, after nearly 14 billion years, enables the Universe to look back and reflect upon itself.

Because the cosmos is a learning system, a primary purpose for being here is to learn from both the pleasures and the pains of existence. If there were no freedom to make mistakes, there would be no pain. If there were no freedom for authentic discovery, there would be no ecstasy. In freedom, we experience both pleasure and pain in the process of developing our identity as beings of both earthly and cosmic dimensions.

We stand upon the Earth as agents of self-reflective and creative action who are engaged in a time of great transition and consciously learning to live in a living universe. An ancient Greek saying speaks directly to our learning journey, "Light your candle before night overtakes you." If the Universe were non-living at its foundation, it would take a miracle to save us from extinction at the time of death, and then to take us from here to a heaven (or promised land) of continuing aliveness. However, if the Universe is alive, then we are already nested and growing within its aliveness.

All things end.
All being continues.
That is the nature of each.

When our physical body dies, the life-stream we are makes its passage to a fitting home in the larger ecology of aliveness. We don't need a miracle to save us—we already exist within the miracle of sustaining aliveness. Instead of being saved from death, our job is to bring mindful attention to the ever-emerging aliveness in the here and now. We are shifting from seeing ourselves as accidental creations wandering through a lifeless cosmos without meaning or purpose to seeing ourselves engaged in a sacred journey of discovery in a living cosmos of stunning depth and richness of purpose.

In learning to live in a living Universe,
we are learning to live in the deep ecology of existence.
This is such an astonishing call to our soulful nature
from the deep compassion of a living Universe
that we would be cosmic fools
to ignore an invitation whose value
is beyond price or measure.

An old saying goes, "A dead man tells no stories." In a similar way, "A dead Universe tells no stories." In contrast, a living Universe is itself a vast story continuously unfolding with countless characters playing out gripping dramas of awakening and creative expression, inseparable from the artistry of world-making. The Universe is a living, unfolding creation. Saint Teresa of Avila saw this when she wrote, "The feeling remains that God is on the journey, too."[201]

If we see ourselves as participants in a cosmic garden of life that has been growing patiently over billions of years, then we feel invited to shift from feelings of indifference, fear and separation to feelings of curiosity, love, and participation.

If we are no more than biological entities, then it makes sense to think we could disconnect ourselves from the suffering of the rest of life. However, if we are all swimming in the same ocean of subtle aliveness, then it is understandable that we each have some measure of direct experience of being in communion with the larger fabric of life—including both the joy and the suffering of all existence. Because we share the same matrix of existence, the totality of life is already touching each of us and co-creating the field of aliveness within which we exist.

Natural Ethics

A felt ethics emerges from our intuitive connection with the living Universe in the form of a "moral tuning fork." We can each tune into the non-local field of life and sense what is in harmony with the wellbeing of the whole. When we are in alignment, we experience a warm, positive hum of wellbeing as a kinesthetic sense that we may call "compassion." In a similar way, we can also experience the dissonant hum of discordance and dissatisfaction.

In recognizing we can contribute with discernment to the unfolding story of cosmic evolution, we shift from a sense of existential disconnection to feelings of intimate communion with all that exists. In turn, we are less inclined to turn to external authorities for our direction regarding what is "right" and "wrong." Instead, we discover the wisdom arising from our experience of aliveness as the guiding compass for our pathway through life. Vandana Shiva, environmental activist and physicist, expresses this insight in practical terms, writing, "In nature's economy the currency is not money, it is life."[202]

Aliveness is our greatest wealth. This wisdom was made clear by Professor Jem Bendell, a climate activist, in an interview with the Buddhist practitioner and spiritual elder, Joanna Macy, as she spoke of our perilous time of transition and adaptation. Joanna offered this connection of climate activism with aliveness:

"The current moment is an exquisite time to be alive. Because, an awareness of impending collapse is an invitation to ask ourselves deep questions of meaning that we typically postpone—and some of us never even get to. *Climate despair is inviting people back to life.* . . . The way through despair involves experiencing oneself as part of a greater whole and surrendering to the mystery of creation. . . . The climate crisis invites us to engage with the mystery of life with fresh eyes and an open heart."[203]

The pull of aliveness is blossoming in the world as a commonly understood experience. Awakening to our conscious connection with the living Universe naturally expands our scope of concern and compassion—and brightens the prospect of working together to build a sustainable future.

8. Waking Up

Despite the harshly difficult scenario of how life is currently unfolding for humanity, there are reasons for optimism. In the previous section, we saw how the world is beginning to awaken to a new regard for aliveness. An intimate companion to aliveness is a growing regard for consciousness. An awakening consciousness brings with it an enhanced capacity for creative choice and is vital for moving through our time of great transition and choosing a regenerative pathway into the future. Let's explore briefly how the further awakening of our knowing faculty—our consciousness—can serve our pathway ahead.

Because consciousness or a knowing capacity is fundamental to our nature as humans, the awakening of consciousness is a natural aspect of our evolution. We can accept our scientific name as *Homo sapiens sapiens*—beings who have the ability to "know that we know" and can watch or witness ourselves as we move through our daily lives. We see that to the extent we are not running on automatic—following habitual and pre-programmed ways of living—we have far greater freedom and choice—an invaluable capacity in this time of great transition.

A unifying insight in all of the world's spiritual traditions as well as in psychotherapy is that the first step in awakening and healing is to simply see "what is"—in other words, becoming an objective witness or impartial observer of our experience. Honest reflection and nonjudgmental witnessing are fundamental to both individual and collective awakening.

In paying attention to our lives in the mirror of consciousness, we can make friends with our soulful nature and come to greater self-possession. The capacity for honest, self-reflection provides a way to cut through the surface chatter of our lives and discover the direct experience of our soul. We can then begin to take soulful responsibility for our lives and relationships.

Peter Dziuban writes and speaks about consciousness and aliveness.[204] He brings clarity to aliveness as a direct experience instead of something we think about. He asks us to imagine a wine tasting party where tasting the wine is what it's all about. The tasting is the purpose. So it is with Life. We are here to taste what it means to be alive—to directly experience and live our aliveness. Dziuban writes, "Life is nothing if it is not alive!" In the simplicity of silence, we can taste aliveness. Our aliveness is not a thought but a living presence. Aliveness is not a thought *about* aliveness, it is being the *actual aliveness itself.*

> "You are conscious and alive. The words and thoughts are what you are *conscious of*. Words and thoughts by themselves are never conscious—only you are. So that's what you really are, this pure consciousness—not unconscious words and thoughts *about* it. Huge difference. Thinking is a changing process. Aliveness is a changeless Presence."[205]

To witness or watch ourselves moving through life is not a mechanical process but a living experience where we are consciously "tasting" our lives, and making friends with ourselves, including those places of doubt, anger, fear, and desire that we may prefer to ignore.

An "observer self" or "witnessing self" provides the ability to stand back from complete identification with bodily-based desires, emotions, and thoughts. With the trustworthy mirror of reflective consciousness, we can see ourselves as if from a distance. From this perspective we see that, while our bodily experience is one part of ourselves, we are more than our body's sensations, pleasures, and pains. We also see that while our emotional experience is one part of ourselves, we are more than our experience of anger, happiness, and sorrow.

By bringing a reflective consciousness into our lives, we have more spaciousness and freedom, as we are no longer identified exclusively with sensations, emotions, and our inner stream of mental dialogue. It is the detachment and perspective provided by a reflective knowing or consciousness that supports the communication and reconciliation vital for moving through our times of great transition.

In being present with a reflective consciousness, we are no longer operating largely on automatic. Expanding to a social scale, seeing ourselves in the mirror of mass media—the internet, television, and other tools of the global nervous system—changes everything. The recognition that we live in a shared field or ecology of consciousness provides the connecting thread to weave the human family into a mutually appreciative whole while simultaneously honoring our differences.

All people share in the common ocean of consciousness. Irrespective of differences in gender, race, wealth, religion and so on, we all participate in the deep ecology of consciousness and this provides a common ground for meeting, understanding, and reconciliation.

Reconciliation does not mean that past injustices and grievances are erased; rather, by being consciously acknowledged and coupled with sincere efforts for restoration, they no longer stand in the way of our collective progress. When injustices are consciously recognized along with public apology and reparation, it releases both parties from the need to continue the process of blaming and feeling resentful; instead both can focus on restorative and cooperative actions for building a constructive future.

A reflective consciousness is vital if we are to cope with intense global stresses and challenges. We have entered the perfect storm of an intertwined system of critical problems and require an unprecedented level of global reflection and reconciliation around a shared vision of a sustainable future.

Importantly, the awakening of consciousness does not end with mindfulness or reflective attention. Beyond a reflective consciousness and the polarity of watcher and watched, or observer and observed, there can grow a unitive consciousness. If we persevere with sustained mindfulness, the distance between observer and observed gradually diminishes until we become a single, integrated flow of experience. As the knower and that which is known converge and become one in experience, we become inseparable from that which we have been observing.

Because the universe is a profoundly unified whole, we are simply allowing our conscious knowing to coincide with that which is being known. We let go of objectifying reality as something to be witnessed "out there" and accept that it can be directly experienced "in here." We move beyond "reflecting on" and move into the experience of "coinciding with."[206] Our direct experience becomes transparent to that which has been the object of our attention.

A new social atmosphere will emerge in a culture of compassionate consciousness. No matter where people are in the world, we will increasingly know we are among relatives. Our sense of identity will expand and we'll regard ourselves as "compassionate citizens of the cosmos"—beings who are immersed within the depths of a living Universe and who feel a deep kinship with all of life.

The word "passion" means "to suffer" and the word "com-passion" literally means "to suffer with." If we are watching the world move through the throes of painful transition, we become one with the ocean of suffering and will work to alleviate that suffering. Swimming in the larger ocean of life, we know intuitively that, if the Earth is suffering, we are all bathing in an ocean of subtle suffering. We recognize that our experience of life is permeable and that we share in whatever measure of happiness or sorrow being created for the whole.

As the push of outer necessity meets the pull of untapped inner capacity, humanity is beginning to awaken. And yet, adversity trends such as radical climate change are accelerating so rapidly there is a real danger that humanity's responses could prove to be too little and too late—and we may veer off into a new dark age. If we are distracted and in denial and overlook the urgency and importance of the great transition now underway, we will miss a unique, never-to-be-repeated, evolutionary opportunity.

Each generation makes sacrifices for the next as a caretaker for the future. The current generation is being pushed by a wounded Earth and pulled by a welcoming Universe to make an unprecedented gift to humanity's future: awakening together with equanimity and maturity to consciously realize our bio-cosmic potential and purpose of learning to live in a living Universe.

9. Growing Up

One of the most insightful ways of understanding humanity's great transition is by viewing our journey through the lens of our level of collective maturity. How grown up are we as a species?[207]

Over the last 30 years, in speaking with audiences around the world about humanity's future, I've often begun my presentation by asking the following question: "When you look at the overall behavior of the human family, what life-stage do you think we are in? If you estimate the social average of human behavior, what stage of development best describes the human family: Toddler? Teenager? Adult? Elder?"

I have presented this same question to diverse business leaders in Brazil, the US and Europe; to spiritual leaders and groups in Japan and the US; to women graduating as school teachers in India; to non-profit groups and student groups in the US, Canada, and Europe; to an international community of women leaders, and so on. Wherever this question has been asked, people are ready to respond within seconds—and the same response has consistently come back from around the world: *Roughly three-quarters say that humanity, taken as a whole, is in its adolescent years!*

The speed and consistency with which diverse people around the world come to this intuitive conclusion has been striking. It is clear that we share a common view regarding the life stage of the human species: We are in our *collective adolescence*. The most common reasons offered for this point of view are:

- Adolescents are often *rebellious* and want to prove their independence. Humanity has been rebelling against nature, trying to demonstrate our independence and superiority.

- Adolescents can be *reckless* and inclined to live without regard for the consequences of their behavior, often feeling they are immortal. The human family has been recklessly consuming natural resources as if they would last forever, polluting the air, water and land, and extincting a significant part of animal and plant life on the Earth.

- Adolescents are frequently concerned with outer *appearance* and with fitting in materially. Many humans are concerned with how they express their identity and status through material possessions.

- Adolescents are inclined toward instant *gratification*. As a species, we are seeking our short-term pleasures and largely ignoring the long-term needs of other species and future generations.

- Adolescents tend to gather in groups or cliques, and often express this as "*in versus out*" thinking and behavior. Much of humanity is clustered into political, socio-economic, racial, religious, and other groupings that separate us from one another, fostering an "us versus them" mentality.

Importantly, some have offered a strongly positive side of adolescence saying, for example, adolescents have a huge amount of energy and enthusiasm and, with their courage and daring, are ready to dive into life and make a difference in the world. Others have said, many adolescents have a "hidden sense of greatness" and feel that, if given a chance, can accomplish remarkable things.

We have yet to step into our early adulthood as a species and, if we liberate ourselves from the constraints of the past, we can find untapped energy, creativity, and courage to achieve a level of greatness that is now hidden.

When I've asked people, what was most important for them personally in moving from their adolescence to adulthood, common themes have emerged that are instructive for humanity's great transition. Often mentioned were:

- A brush with death where a friend or family member died and they were reminded of our mortality as humans and how we have a limited time on this Earth to learn and grow. People often mentioned how humanity is facing the threat of species extinction and this could be a powerful push into our early adulthood.

- Role models and how a person inspired them to reach beyond their current scope of behaviors and explore new potentials for themselves. Many also noted that current role models celebrated in the mass media tend to be movie stars, sports stars, and popular musicians. People recognized that these role models encourage adolescent behaviors and did not draw them into their early maturity.

- Another common theme was being pushed to take responsibility for others and to move beyond typical adolescent concerns. For example, the need to care for a sibling, an aging parent or a sick friend, or to take an extra job to earn money needed by the family. People recognized that we are being pushed to take responsibility for the well-being of the Earth as we move into a time of great transition.

- Others said they took a "hard look in the mirror" and saw how they were conducting their lives in adolescent ways—and this reflective feedback gave them the motivation to shift toward early adulthood. As a human family, the internet and television are giving us a penetrating view of ourselves and the consequences of our behavior and the need to step up to a higher level of maturity.

If people around the world are accurate in their overwhelming view that the human family is now in its adolescence, it explains much of our current behavior—and it suggests how we could behave differently as we move collectively into our early adulthood. Here are six important shifts we can choose to make in how we show up in the world and that demonstrate greater maturity:

- **Adults are able to give priority to others before themselves.** With greater maturity, adults are able to look beyond self-centered wants and desires and, instead, consider how they can serve the well-being of others and the Earth. Rather than being self-absorbed, adults can be selfless and make sacrifices for others without feeling resentful. A mature person and society can find joy in the success of others and derive satisfaction from sharing their good fortune with others.

- **Adults are able to keep long-term commitments and choose delayed gratification.** If we are to make a commitment to future generations and pull back from the current path of ecological catastrophe, it will be essential for the materially privileged of the world to pull back from a consumerist mindset and insure there will be sufficient resources and a habitable planet for future generations. More than token generosity, the entire global economy needs to be reconfigured for equity and common good. This is truly an undertaking for mature adults.

- **Adults have a greater sense of humility.** Adults move beyond shallow stereotypes and are less concerned with their self-image being governed by how many things they own or how they look in comparison with others. Adults do not feel they have to prove themselves to others with disproportionate material possessions; instead, they can choose more modest and unpretentious ways of living. A mature individual and society do not regard themselves as "special" and deserving of material privileges, instead, they tend believe in fairness and the equal rights of others.

- **Adults tend to be more self-accepting and accepting of others.** A mature person or society has been seasoned by life experience and recognizes we are here for more than pleasure seeking—we are here to learn, grow, and give—and this inevitably involves making mistakes. In our maturity, we accept our humanity and have greater compassion for ourselves and others.

- **Adults tend to talk less and listen more.** A mature person will tend to listen for understanding rather than listen for interruption to make an argument for their point of view. It is vitally important for mature leaders to listen deeply to people—especially the young and marginalized—and to reflect back to them what is heard. Adult leadership recognizes that listening and learning go together—both are invaluable skills for a world in great transition.

- **Adults take charge of cleaning up after themselves.** Adults don't expect others to clean up the messes they make. Instead of waiting for others to handle things, adults take charge of how our lives impact others. Instead of waiting for a mythical "mom or dad" to fix the climate crisis, a mature person or society listens and brings focused attention to a troubling situation and then acts, recognizing that our personal and collective lives are deeply intertwined.

In the coming decades, the human family will be pressed to take a hard look in the "social mirror" of mass media and recognize we are behaving like adolescents in a world that requires adults if we are to make the transition to a workable future. A new maturity is the basis for a new economy as adults shift from a focus on material obsessions and possessions to soulful development of ways of living that bring greater meaning, aliveness, and awareness to the forefront of our lives. It is vital that the human family also shift other priorities:

- From division to reconciliation and working to heal the separations of race, gender, sexual orientation, age, politics, geography, class, and religion and find ways to live and work together productively as a human family. With greater maturity, those among us who have more privilege recognize the deep wounds of history and extend humble support and offer compensation to those who have experienced great injustice and oppression, for example, by being enslaved or having their ancestral lands taken away.

  We who have greater privilege also recognize and seek justice for those who are economically marginalized or discriminated against or disenfranchised because of their race, gender, class, and other identities. This includes support for generating economic opportunity or for relocating as refugees and adapting as quickly as can those who are in more privileged circumstances.[208]

- From reckless consumption of resources to a conscious stewardship of the Earth, recognizing countless generations to come will call this planet their home. In turn, moving from consumer-oriented nations focused on instant gratification and economic growth at all costs to finding satisfaction in wise, restrained consumption and the knowledge that a sustainable world is being established for future generations.

- From a human-centered perspective that enables the mass extinction of animal and plant life on Earth to a life-centered perspective that gives the highest priority to ensuring we have a resilient diversity of species.

- From rebelling against nature to designing ourselves back into nature as an integral part of the natural world; for example, developing new forms of housing and common spaces that foster stronger communities as a social and physical architecture for sustainable living.

- From a feeling of immortality to feelings of vulnerability and humility in recognizing the rise and fall of civilizations throughout human history and that we are approaching a world-systems crisis that threatens the viability of many civilizations and ecosystems.

- From passivity and tolerance of exploitative media to seeking accountability for serving the well-being of life with programming that informs people about the actual conditions of the world.

- From a politics of wanting "mom or dad" to fix things (whether big government or big business), to a politics of greater self-reliance at the local level. Strengthened community and democracy at the local level supports new levels of civic engagement at the national and global level, particularly by mobilizing the power of internet-based communication.

Growing up is entirely natural—and we can do this both as individuals and as an entire human family. Yet, we should not diminish how truly demanding this journey is: Maya Angelou wrote these powerful lines describing the difficulty of growing up:

> "I am convinced that most people do not grow up. We find parking spaces and honor our credit cards. We marry and dare to have children and call that growing up. I think what we do is mostly grow old. We carry accumulation of years in our bodies and on our faces, but generally our real selves, the children inside, are still innocent and shy as magnolias."[209]

Toni Morrison said in a commencement speech, "True adulthood is a difficult beauty, an intensely hard won glory, which commercial forces and cultural vapidity should not be permitted to deprive you of."[210] Moving beyond our adolescence and into our early adulthood, the human community could give rise to a dramatic transformation in our approach to living on the Earth in these transitional times.

10. Reconciling

Only together can we realize a great transition. Transition is a team effort—all hands, on deck! A team effort is impossible if we are deeply divided as a human community. The world is awash with racial and gender discrimination, genocide, religious wars, oppression of ethnic minorities, and the extinction of other species. Some of these tragedies have grown and festered over thousands of years and this makes pulling together in common effort very difficult. Nonetheless without deep and authentic reconciliation across these and other barriers of division and suffering, humanity will remain separated and mistrustful—and our collective future will be gravely imperiled.[211]

As difficult and uncomfortable as this process might be, conscious reconciliation that includes truth telling, public apology and meaningful reparation is a vital part of our collective healing and is essential if humanity is to move forward together on our journey. The Earth community faces a stark choice for the future—do we:

- **Pull together** as a human community, accepting all of the *sacrifices* that will be involved, or

- **Pull apart** as human sub-groups, enduring all of the *violence* that will inevitably result.

If the people of the Earth are able to pull together and collaborate for well-being for all, a cascade of actions and innovations can follow quickly from the clarity of our unified, social will. However, if the social will of the people of the Earth is not awakened on behalf of our *collective* well-being but remains deeply divided, then we seem likely to turn toward either the seeming safety of authoritarianism or fragment into countless sub-groups as unresolved wounds and divisions continue and produce ever deeper separation and increased violence.

A world divided against itself is a recipe for global collapse and humanity's functional extinction. We can acknowledge the wisdom of Dr. Martin Luther King, Jr.: "We must learn to live together as brothers or perish together as fools."[212] In the words of Alan Paton, "It is not 'forgive and forget' as if nothing wrong had ever happened, but 'forgive and go forward,' building on the mistakes of the past and the energy generated by reconciliation to create a new future."[213]

Although we can see the broad outlines of a sustainable future, the human family is far from being ready to work together. To come together, the Earth family must engage in a process of authentic reconciliation across a number of areas; for example:

- **Gender, racial, sexual and ethnic reconciliation**—Discrimination profoundly divides humanity against itself. To work together for our common future, we must build a global culture of mutual respect that enables us to work together as equals. This does not mean we will ignore gender, racial, sexual, and ethnic differences; rather we learn to respect and include differences, and then work to transform oppressive structures and systems. We move beyond limiting judgements of others, especially those that do not fit the dominant culture and, instead, weave a new culture that is respectful, inclusive, and fair.

- **Generational reconciliation**—Sustainable development has been described as that which meets the needs of the present without compromising the ability of future generations to meet their needs.[214] Because many industrial nations are using up vital nonrenewable resources in the short run, it severely limits the options of future generations to meet their needs. To pull together, we must reconcile ourselves across generations. For example, adults can support youth by listening to their needs, shining a light on youth movements and concerns, and hearing how the lives of adults have helped to create the climate crisis.

- **Economic reconciliation**—Enormous disparities exist between the rich and the poor. Reconciliation requires narrowing these differences and establishing a world, minimum standard for economic well-being that supports people in realizing their potentials. Yale professor, Narasimha Rao states, ". . . reducing inequality—within countries and between them—would improve our ability to mitigate some of the worst effects of climate change, and provide for a more stable climate future. . . climate change, at its most essential, is a justice issue."[215] United Nations research reveals that global inequality is often more about disparities in opportunity than disparities in income.[216] Perhaps the deepest change may be to disconnect how personal value is associated with one's position in a hierarchy of wealth or social class.

- **Ecological reconciliation**—Living in sacred harmony with the Earth's biosphere is essential if we are to survive and evolve as a species. Restoration of the biosphere is vital as our common future depends on the presence of a broad diversity of plants and animals. To move from indifference and exploitation to reverential stewardship will require reconciliation with the larger community of life on Earth, and honoring those who have been preserving cultures of sacred reciprocity with all life.

Consumer cultures are placing human needs above those of the Earth community and this has led to ecological disasters. We humans are an inseparable part of the Earth and what happens to the Earth happens to us. We can protect the wilderness areas that remain and bring genuine concern for the well-being of the living world into all we do.

- **Religious reconciliation**—Religious intolerance has produced some of the bloodiest wars in human history. Vital to humanity's future is reconciliation among the world's spiritual traditions (for example, Catholics and Protestants in Northern Ireland, Arabs and Jews in the Middle East, Muslims and Hindus in India). As the world's religious and spiritual traditions become more accessible through the internet and communications, we can learn to appreciate the core insights of each tradition and see each as a different facet in the common jewel of human spiritual wisdom.

Many of these divisions are starkly evident in our world and, with climate disruption, will disproportionately impact women and the world's poor most profoundly. Here is a compelling summary from a recent Oxfam briefing:

"Within countries it is often the poorest communities—and particularly women—who are most vulnerable. Poor communities tend to live in poorly built houses on marginal land that is more at risk from extreme weather such as storms or floods. They often live in areas with poor infrastructure, making it difficult to access essential services such as healthcare or education in the aftermath of an emergency. They are unlikely to have insurance or savings to help them rebuild their lives after a disaster. And many depend on farming or fishing—activities which are particularly vulnerable to more extreme and erratic weather. With the frequency and intensity of climate-related hazards increasing, the ability of people living in poverty to withstand shocks is gradually being eroded. Each disaster is leading them in a downward spiral of deeper poverty and hunger, and eventually displacement. . . ."

"Women are often among the last to leave home when more extreme or erratic weather makes it harder for families to put food on the table, staying behind to look after children and elderly or sick relatives, while male family members leave to search for an income elsewhere. This can place a huge burden on women, who often become the main provider for the family as well as the primary caregiver. Their jobs are made harder by the climate crisis, which makes growing food and collecting water and fuel more difficult and time-consuming. . . ."

"When forced to leave home, women and children are particularly vulnerable to violence and abuse. . . . Displaced children are often denied an education, locking them in an inter-generational cycle of poverty. Gender inequalities also make it harder for displaced women to rebuild their lives. . . . Lower education and literacy rates mean displaced women often lacked information on their legal rights which could help secure their access to land."[217]

Great personal and social maturity will be required to recognize and remedy injustices and injuries so that the human family can work together for our common well-being. Bringing legitimate grievances into public awareness, mourning the mistakes of the past, taking responsibility for them, and then seeking just and realistic remedies—these difficult actions are at the heart of the era of reconciliation.

We require unprecedented communication
to discover our common humanity
from a place of uncommon humility.

With reconciliation and restoration, social energy that was previously locked up in oppression and injustice can be freed up and become available for productive working relationships.

To move ahead, we must find ways to bridge the past with the future—to recognize, reconcile and remedy the separations and wounds of the past with the inevitable sorrows, losses and traumas of the future. Can we rise to a higher maturity to inhabit this unfolding stream of time and find ongoing reconciliation and restoration?

The process of reconciliation is complex and involves three major steps: the injured must be heard publicly, the wrongdoers must apologize publicly, taking responsibility for the impacts of their choices, and then they must provide restitution or reparations that make amends for the past, and provide a foundation of greater wholeness for all to move into the future together.

Being heard is the first step in being healed. By listening to and acknowledging the stories of those who have suffered, we begin the process of healing. Our collective listening to the wounds of humanity's psyche and soul is vital to our collective healing. Listening does not mean forgetting; instead, it means to bring the wounds of division into collective awareness and to remember these as we seek ways to move into the future.

Archbishop Desmond Tutu knows more about the process of reconciliation than most. He was the chairman of the Truth and Reconciliation Commission (TRC), which was established to investigate crimes committed during the apartheid era in South Africa in the period from 1960 to 1994. When apartheid ended, South Africa's black majority had to choose among three different ways to seek justice and live together with the country's white minority. They could choose justice based on *retribution*—an eye for an eye; or justice based on *forgetting*—don't think about the past, just move forward into the future; or justice based on *restoration*—granting amnesty in exchange for truth. Archbishop Tutu explains their choice:

We believe in restorative justice. In South Africa, we are trying to find our way toward healing and the restoration of harmony within our communities. If retributive justice is all you seek through the letter of the law, you are history. You will never know stability. You need something beyond reprisal. You need forgiveness.[218]

A second step in being healed is for the wrongdoer to offer a sincere public apology. Here are examples of important public apologies:[219]

- **In 1988**, an act of Congress apologized "on behalf of the people of the United States" for the internment of Japanese Americans during World War II.

- **In 1996**, German officials apologized for the invasion of Czechoslovakia in 1938 and established a fund for the reparation of Czech victims of Nazi abuses.

- **In 1998**, the Japanese prime minister expressed "deep remorse" for Japan's treatment of British prisoners during World War II.

- **In 2008**, the US Congress formally apologized for the country's "original sin"—its treatment of African Americans during the era of slavery and subsequent laws that discriminated against blacks as second-class citizens in American society.

Another powerful example of a public apology and social healing is provided by the relationship between the Aboriginal people and the European settlers in Australia. In 1998, Australia commemorated its first "Sorry Day" to express people's regret and shared grief about a tragic episode in Australian history—the organized removal of Aboriginal children from their families on the basis of race.

Through much of the 20th century, Aboriginal children were forcibly removed from their families with the aim of assimilating them into Western culture.[220] Sorry Day provides a way for Australians to come to terms with their history and to remember together as a way to build a future on a foundation of mutual respect. Indigenous council member, Patricia Thompson stated, "What we want is recognition, understanding, respect and tolerance—of each other, by each other, for each other." In cities, towns, and rural centers, in schools and churches, people stop their everyday activities to acknowledge this injustice. In addition, hundreds of thousands of

Australians have signed the "Sorry Books." An essential require-ment for reconciliation is conscious requests for forgiveness as well as remembering.

The third step in reconciliation is restitution or the payment of reparations. Archbishop Desmond Tutu explains the role of res-titution when he says that reconciliation involves more than the recognition and remembering of injustice: "If you steal my pen and say 'I'm sorry' without returning the pen, your apology means nothing."[221] What is also needed is restitution. Apologies create a truthful record. Restitution creates a new record. The purpose of reparation is to repair the material conditions of a group so as to restore balance or equality of power and material opportunity.[222]

With authentic reconciliation—that includes listening, remem-bering, apologizing, and restoring—the divisions and suffering of the past do not need to stand in the way of future harmony. This is not as simple as providing money or land or policies designed to remove inequities. The deep wounding of the oppressed also man-ifests as generational trauma that no amount of money will erase. True reparation must provide for healing and wholeness.

As difficult and uncomfortable as this process will be, this is a vital stage in our collective healing that can enable humanity to move forward on our common journey. As our planetary consciousness awakens, deep psychic wounds and traumas will emerge that have remained hidden through history. We will begin to hear the voices that have been ignored and the pain that has not been acknowledged. Awakening will bring into collective awareness the dark shadows of human history, including racism, ethnic conflict, religious persecu-tion, gender oppression, and economic discrimination.

Just as a rising tide lifts all boats, so too can a rising level of global communication lift all injustices into the healing light of public awareness. Our ability to communicate with ourselves as a planetary species about these painful wounds will be critical for reconciliation.

11. Communicating

Many of today's largest institutions have two, profound limitations: They embody a restrictive materialistic paradigm and have grown to overwhelming complexity. In turn, many institutions have neither the worldview nor the organizational capacity to evolve with the speed and perspective required by our world in great transition. In addition, most citizens in technologically advanced societies woefully lack an understanding of the urgency and severity of trends such as the climate crisis. We are unprepared people in unprepared societies. We can begin to see this by looking at the dominant media in "advanced" nations.

If we use the United States as an example, we see how extremely inadequate the level of attention is to climate changes as presented through broadcast television, which continues to be a primary source of information for a majority of citizens, despite the growing presence of social media in our lives. If we combine the number of minutes of climate coverage for the broadcast TV networks (ABC, CBS, NBC, and Fox), we see that the total number of minutes for an entire year of news programming dropped from just over 4 hours in 2017 to just over 2 hours in 2018. *This is an astoundingly inadequate level of attention to climate change for a modern democracy facing a planetary crisis!*

Figure 8: Combined Minutes of News Coverage of Climate Change on US Broadcast TV: 2017 & 2018[223]

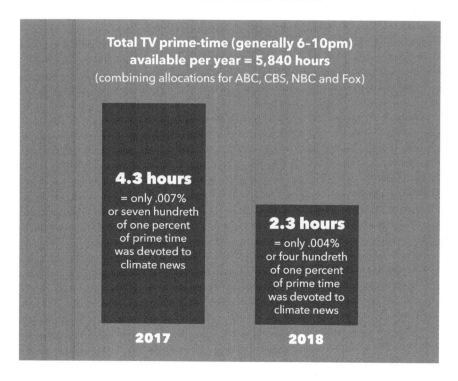

If we look back another year to the critical election year of 2016 in the US, the pattern is even more clear: For the entire year of 2016, broadcast TV coverage of climate change for the evening news for all four of these huge, national networks combined was only 39 minutes.[224] Less than an hour of coverage for an entire year! This is such a diminished level of attention that it illustrates, with startling clarity how, in service of corporate profits, the US is being devastatingly dumbed-down as a society.

Drawing lessons from the example of the US, how can the people of the Earth move beyond debilitating impoverishment of our collective awareness and understanding? By learning together and communicating at entirely new levels, the citizens of Earth can become a creative team for transformational change. Using the new media made possible with the internet, cell phones, computers and more, we can step above aging institutions to bring a stunning new source of inspiration and influence that encourages rapid evolution.

It was our ability to communicate that enabled us to evolve from awakening hunter-gatherers to the threshold of a planetary civilization. In turn, by mobilizing the power of local to global communication we can step into our early maturity and the beginning stage of a sustainable species-civilization. To shift toward building a regenerative future will require the voluntary choices and actions of billions of people.

Never before in human history
have so many people been called upon
to make such sweeping changes in so little time.

To act with the speed, cooperation and creativity demanded by our situation, we will require the respectful participation of free individuals consciously acting in concert with one another. People will resist making changes unless—and until—we have reached broad agreement about the nature of change that is essential.

As a first step, people will need to communicate their despair that the global ecology cannot be restored, their resentments for broken dreams of material prosperity, and their unwillingness to make sacrifices. After a cathartic process of learning, communication, reconciliation and restoration, people will be ready to act with the level of energy, creativity, and cooperation our circumstances demand.

However, once citizens know what other citizens locally—and around the world—are willing to do, and once they are settled in their own hearts and minds as to what constitutes appropriate action, then they and their representatives in government can act swiftly and with authority.

Communication is the lifeblood of democracy and civilization. A mature democracy and society require the active participation and consent of an informed public, not simply their passive acquiescence. Democracy has often been called the art of the possible. If we don't know how our fellow citizens think and feel about policies to create a sustainable future, then we float powerlessly in a sea of ambiguity—unable to mobilize ourselves in constructive action.

The most powerful and direct way to revitalize democracy is by improving the ability of people to know their own minds at every scale—local, national, and global. By combining televised "Electronic Town Meetings" on key issues with instantaneous internet-based feedback from a scientifically selected sample of citizens, the public can know its collective sentiments with a high degree of accuracy. With regular Electronic Town Meetings (or ETMs), the perspectives and priorities of the citizenry can be rapidly brought into public view and the democratic process revitalized. When a working consensus emerges, it can guide decision makers.

The value and purpose of ETMs is not to micromanage government through direct democracy; rather, it is for citizens to discover their widely shared concerns and priorities that can guide their representatives in government. Involving citizens in choosing the pathway into the future will not guarantee that the "right" choices will always be made, but it will guarantee that citizens will be involved and invested in those choices. Rather than feeling cynical and powerless, citizens will feel engaged and responsible for our future.

At the local level, a non-partisan and non-profit "Community Voice" organization can work in cooperation with metropolitan scale television broadcasters and internet providers to produce an ETM with feedback from a scientifically selected or random sample of citizens. This can be supplemented with feedback from specific groups (younger, older, of diverse ethnicities, genders, religions, etc.) that want to participate in non-random surveys via the internet.

An entirely new avenue of citizen learning and engagement beckons. To illustrate, the "Extinction Rebellion" is an international movement that uses non-violent, civil disobedience in an attempt to gain greater media attention for halting mass extinction and minimizing the risk of civilizational collapse.[225] ETMs are an alternative approach that go directly to the need for citizen dialogues by creating independent, non-partisan, metropolitan-scale organizations that can serve as respected vehicles for developing a new level of public discourse. (ETMs that can operate from the local to global scale are explored in greater length in the Appendix.)

Looking ahead: As "soft denial" and willful ignorance of climate disruption shifts to recognition of the profound seriousness of our global situation, entirely new levels of collective attention and communication will be essential if we are to find our way—together—into a workable and meaningful future.

Fortunately, for the first time in history, a majority of the people of Earth already have the tools to engage in a collective conversation regarding our pathway into the future. In 2019, nearly 60% of the global population had internet access.[226] More than half of the world's population with TV sets are now within reach of a digital TV signal.[227] Extrapolating from these trends, roughly 75% of global population will have internet access by the end of the 2020s.[228]

Although a technological revolution in communications has been underway for at least three decades, socially, humanity has yet to recognize the astonishing power of these technologies and to mobilize ourselves for communicating our way into a workable and purposeful future.

The next great superpower will not be a nation
or even a collection of nations;
rather, it will be the billions of ordinary citizens
who encircle the Earth and who call, with a collective voice,
for unprecedented cooperation and creative action
to care for our endangered Earth and for humanity
to grow into a mature, planetary civilization.

This new superpower is emerging from the combined voice and conscience of the world's citizens mobilized through a global communications revolution. When the people of Earth are more than passive recipients of information (as *witnesses* to climate disruption, intense poverty, species extinction, and more), but also are able to offer our collective *voice* for change, then a new and powerful force for creative transformation will be unleashed in the world.

A genuine "Earth Voice" movement must be able to reach the vast majority of people virtually all at the same time. Seven billion or more people need to be able to offer their feedback about critical concerns if they are to feel part of an Earth Voice conversation. This level of functioning of our "global nervous system" would have been completely unthinkable prior to the emergence of the internet that now includes 60% of the world's population—and is growing rapidly.

In the decade of the 2020s, the world will have the technologies with which to acquire real-time feedback and knowledge of humanity's sentiments and views—and humanity will move into a new era of collective transparency, self-reflection and self-discovery. This could develop into non-partisan, global conversations with democratic gatherings of hundreds of millions—or even billions—of people joining in the dialogues. In this way, the collective intelligence of our global nervous system could take on a life of its own in a process that transcends artificial boundaries of communities and nations.

We are vastly over-endowed with all the communication tools we need. All that is required is to step up—from the local to global scale—and make use of these powerful tools of communication and social transformation.

Oscar Morales, a citizen of Colombia, provides an example of digital empowerment and the untapped potential for the future. In 2008, Facebook was growing rapidly in Colombia and Morales began to explore the power of the new social media. He was angry with the terrorism and violence in politics and was so infuriated with the FARC rebels (The Revolutionary Armed Forces of Colombia) of his native country that he decided to express himself.

Late in the evening of January 4th, 2008 he created a Facebook page and named it "One million voices against FARC" that was dedicated to bringing down FARC and to demanding the release of several hundred hostages. By 9 a.m. the next morning, he found that fifteen hundred people had already joined his group. By late afternoon the group had grown to four thousand members. By the second day, the group not only had eight thousand members, but people were actively posting on the discussion board and seeking to connect with him physically and publicly.

As a result of his catalytic posting, on February 4, 2008—a mere four weeks after the group was begun—millions of Colombians marched throughout the country, and in major cities worldwide, to express their anger at FARC. In the space of a single month, one individual had catalyzed millions of people to come together in 27 cities in Colombia and 104 other cities around the world to march in empathy and solidarity.

An Earth Voice movement requires that a majority of people recognize how the mass media is being misused to promote sensationalist advertising that pollutes the public mind, promotes superficiality, and leaves little room for adult reflection. A movement for a more mature media that respects the public's need to be informed and involved in choosing our pathway into the future requires the public to hold mass media accountable and answerable to the public they have an obligation to serve. There are generations of experience with an abusive and exploitative media. The public can rightfully demand media accountability for serving the compelling needs of the public in a time of planetary crisis.

Our evolutionary maturity is being tested. Our future as a species will depend on a new "politics of consciousness" that holds mass media accountable for being a fair witness and mature partner in our collective awakening. By taking charge of our media as mature citizens, we can become self-directing agents of our evolution.

In a democracy, when we are informed as individual citizens, then we "know." However, when we communicate and reflect among ourselves as citizens—publicly learning about and affirming our shared sentiments as an extended community—then we "know that we know." In our dangerous and difficult time of global transition, it is not sufficient for us to be individually knowing or wise; we must become "doubly wise" through social communication that clearly reveals our collective knowing to ourselves.

Once there is a capacity for sustained and authentic social reflection, we then have the means to achieve a shared understanding and a working consensus regarding appropriate actions for a positive future. Actions can then come quickly and voluntarily. We can mobilize ourselves purposefully, and each person can contribute his or her unique talents to building a life-affirming future.

12. Building Community

In our world of great separation, building a strong sense of community is vital. Innovations in the physical design of communities are especially vital for transforming how we live on the Earth. Patterns of living that prioritize sprawling suburbs with isolated households are not well-suited for sustainability. These hyper-individualized growth patterns create formidable barriers to innovation in the future. Growth creates a pattern of living—such as suburban communities—and once these physical forms are established, they then limit the ability to create new patterns of living.

However, with dire necessity as the catalyst for innovation, the 2020s could become a design and retrofitting decade as people create new configurations more adapted to severe ecological and economic limitations. With new zoning rules supporting new forms of infrastructure and with intense pressure to re-invent the global economy, a spectrum of nested innovations could grow from the local to global level:

- **Pocket Neighborhoods** generally consist of a few homes linked together to promote a close-knit sense of community and neighborliness with an increased level of contact.

  "Pocket neighborhoods are generally clustered groups of neighboring houses or apartments gathered around a shared open space—a garden courtyard, a pedestrian street, a series of joined backyards, or a reclaimed alley—all of which have a clear sense of territory and shared stewardship. They can be in urban, suburban or rural areas. A pocket neighborhood is *not* the wider neighborhood of several hundred households and network of streets, but a realm of a dozen or so neighbors who interact on a daily basis around a shared garden, quiet street or alley—a kind of secluded neighborhood within a neighborhood."[229]

- **Eco-Villages** are either designed freshly or, more commonly, retrofitted to provide an integrated way of life with roughly a few hundred or so people. Ecovillages are generally intentional communities, united by shared values and whose goal is to become more socially, culturally, economically and ecologically sustainable. Often, they are locally owned and governed by participatory processes with an intention to regenerate and restore their social and natural environments.

Common to many of these "villages" or co-housing communities is a child-care facility and play area; a common house for community meetings, celebrations, and regular meals together; an organic community garden; a recycling and composting area; a renewable energy micro-grid; a bit of open space for community gatherings; perhaps a play space and conversation space for teenagers; and a workshop with tools for arts, crafts, and repair.

Each micro-community includes diverse types of work as community members trade hours to create a local economy with offerings such as health care, child care, elder care, gardening and permaculture, green building, conflict resolution, internet and electronic support, food preparation, and other skills that provide fulfilling employment for many.

Micro-communities have the culture and cohesiveness of a small town and the sophistication of a larger city as nearly everyone is connected to the world with a rich array of electronic communication. Eco-villages craft unique expressions of sustainability as they create meaningful work, raise healthy children, celebrate life in community with others, and live in a way that seeks to honor the Earth and future generations. The flowering of diverse eco-villages—most created through retrofitting—replace the alienation of today's massive cities.[230]

- **Transition Towns** aggregate neighborhood and village scales into a town with several thousand people who are collaborating at that scale. They often support grassroots community projects that aim to increase self-sufficiency and reduce the potential effects of climate destruction and economic instability. The "Transition Network," founded in 2006, has inspired the creation of many of the transition initiatives that have been started in locations around the world, with many located in the United Kingdom, Europe, North America and Australia.[231]

- **Sustainable Cities** aggregate smaller scales of living into an economic and ecological system of sustainability. "An Ecocity is a human settlement modeled on the self-sustaining resilient structure and function of natural ecosystems. The ecocity provides healthy abundance to its inhabitants without consuming more (renewable) resources than it produces, without producing more waste than it can assimilate, and without being toxic to itself or neighboring ecosystems."[232] Its inhabitants choose ecological lifestyles and the social order seeks to embody fundamental principles of fairness, justice and reasonable equity.

- **Eco-Civilizations** are nations, clusters of nations, and could even characterize the entire Earth community. Eco-civilizations respond to global climate disruption and social injustices with changes that represent an alternative approach to living based on ecological principles. Broadly construed, ecological civilization involves a synthesis of economic, educational, political, agricultural, and other societal reforms toward sustainability.[233]

This nested spectrum of innovation in housing, economic activity and ecological ways of living illustrates how we are beginning to reconfigure our lives to adapt to new global realities. The urgency of shifting to a zero carbon, world economy by 2050 pushes humanity away from the "ego economy" that is ruining the Earth and toward an "aliveness economy" where people bring diverse skills for living simply in communities of caring and purpose. An aliveness economy offers the possibility of both survival and well-being. Many people are learning diverse, new skills for participating in local living economies.

The urban landscape is being retrofitted across a spectrum of scales—from the smallest scale of pocket neighborhoods to the scale of entire eco-civilizations. The eco-village scale is small enough for everyone to know one another and yet large enough to contain diverse skills for creating a micro-economy that serves the many needs of the community (health care, education for new skills, gardens for food, information systems, and much more).

In this century, millions of innovative forms of sustainable and satisfying modes of living are surely going to be developed. Alternative "communities" of every imaginable design will be adapted to local conditions and provide islands of sustainability, relative security, and mutual support via intentional connections with one another in an increasingly chaotic world.

Yet, the strength of local, self-organizing communities can be a weakness if they are seen as isolated havens of safety to weather the storms of transition.

Lifeboats won't save us
when the entire Earth is sinking
and becoming inhospitable to
the biology and psychology of human life.

It is important for the cohesion realized through local collaborations of mutual support to be able to reach widely and provide the social glue to hold larger networks together.

Synergies among pocket neighborhoods and local eco-villages can move up the scale to transition towns and sustainable cities, and then reach to the scale of the world as an eco-civilization.

13. Choosing Simplicity

The magnitude and speed of climate disruption now underway is astonishing and will require dramatic changes in how we live on the Earth—particularly in consumer-oriented societies.[234] For the past few hundred years, consumer-oriented societies have exploited the resources of the Earth primarily for the sole benefit of human beings. This self-serving approach was popular for a time but is now bringing ruin to the Earth. Importantly, the goal of shopping-oriented societies has been to satisfy people's material *wants*; not to satisfy the *needs* for a livable Earth.

Now these economies are being called to reverse a self-serving approach. Instead of asking what we humans *want* (what do we desire, crave, hunger for), we are being called to respond to a far more important question: what does the overall ecology of life *need* (what is essential, basic, necessary) to build a regenerative future for the Earth? A new question is becoming paramount:

What will support an enduring ecology of life,
rather than temporary human comfort?

When we ask, "What can we do to support the ecology of life?" the first powerful action we can take is to bring our personal lives into alignment with the regenerative needs of the Earth. In turn, recall the study, "Life Beyond Growth," which estimated that, "a country like Japan would have to cut its consumption of resources and environmental impact by (very roughly speaking) more than 50%, while the United States would need to reduce by a factor of 75%."[235]

To live sustainably many people will have to choose to live much more simply. At the same time, it is very important to acknowledge that a majority of persons on the planet are already living at the margins of material existence and, for them, simplicity is involuntary—there is no room for meaningful choice in their daily struggles for survival.

Although simplicity is intensely relevant for building a workable world, this approach to living is not a new idea. Simplicity has deep roots in history and finds expression in all of the world's wisdom traditions. More than two thousand years ago, in the same historical period that Christians were saying "Give me neither poverty nor wealth," (Proverbs 30:8), the Taoists were asserting "He who knows he has enough is rich" (Lao Tzu), Plato and Aristotle were proclaiming the importance of the "golden mean"—a path through life with neither excess nor deficit—and the Buddhists were encouraging a "middle way" between poverty and mindless accumulation.

Clearly, the wisdom of simplicity is not a new revelation.[236] What is new is our world pressing against limits to material growth and recognizing the importance of building a new relationship with the material aspects of life.

Simplicity is not opposed to materialism; instead, it places material consumption in a larger context with a different sense of proportion for our lives.

Simplicity does not mean turning away
from either matter or progress; to the contrary,
a progressing relationship with the material side of life
is at the heart a maturing civilization.

Arnold Toynbee—a renowned world historian who invested a lifetime in studying the rise and fall of civilizations—summarized the essence of a civilization's growth in what he called "The Law of Progressive Simplification." [237] He wrote that a civilization's progress was not to be measured in its conquest of land and people;

instead, the true measure of growth is a civilization's ability to transfer increasing amounts of energy and attention from the material side of life to the non-material side—areas such as personal growth, family relationships, time with nature, psychological maturity, spiritual exploration, cultural and artistic expression, and strengthening democracy and citizenship.

Recall that modern physics recognizes that 96% of the known universe is invisible and non-material. The material aspect (including galaxies, stars and planets and our biological being) comprises only 4% of the known universe. If we apply these proportions to our lives, then it is fitting to give substantial attention to the invisible aspects that contain 96% of the known universe and involve often ignored aspects of our lives—the very aspects that Toynbee describes as expressing our progress as a civilization.

Toynbee also coined the word "etherialization" to describe the process whereby humans learn to accomplish the same, or even greater, results using less time, material resources, and energy. Buckminster Fuller called this process "ephemeralization," although his emphasis was on realizing greater material performance for less time, weight, and energy invested. Drawing from the insights of Toynbee and Fuller, we can redefine progress as a two-fold process involving the simultaneous refinement of both the material and non-material sides of life.

With progressive simplification,
the material side of life grows lighter,
less burdensome, more easeful, elegant and effortless
and, at the same time, the non-material side of life
becomes more vital, expressive, and artistic.

Simplicity involves the co-evolution of both inner and outer aspects of life. Simplicity does not negate the material side of life but rather calls forth a new partnership where the material and the non-material aspects of life co-evolve in concert with one another. Outer aspects include the basics such as housing, transportation, food production, and energy generation. Inner aspects include learning the skills of touching the world ever more lightly and lovingly—ourselves, our relationships, our work, and our passage through life.

By refining both outer and inner aspects of life—outward simplicity combined with inner richness—we can foster genuine progress and build a sustainable *and* meaningful world for billions of people without devastating the ecology of the Earth.

An ethic of moderation and "enough" will grow in importance as global communications reveal vast inequities in material well-being. Economic justice does not require replicating the industrial-era manner of living around the world; instead it means every person has a right to a fair share of the world's wealth, adequate to ensure a "decent" standard of living—enough food, shelter, education and health care to be considered sufficient by a reasonable standard of human decency.[238] Given intelligent designs for living lightly and simply, a decent standard and manner of living could vary significantly depending on local customs, ecology, resources, and climate.

To accomplish a great transition within a few decades requires that we invent new approaches to living that transform every facet of life—the work we do, the communities and homes in which we live, the food we eat, the transportation we use, the clothes we wear, the symbols of status that shape our consumption patterns, and so on. We can call this way of living "voluntary simplicity" or "conscious simplicity" or "ecological living."[239] However described, we need more than a change in our style of life.

A change in *style* implies a superficial or exterior change—a new fad, craze, or fashion. We require a far deeper change in our *way* of life, one that recognizes the Earth is our home and must be maintained for the long-range future. Ecological living begins with the understanding that we all live in mutual contingency and that we create safety, comfort, and compassion in our lives together.

An ecologically conscious economy will shift its emphasis from sheer physical expansion to more qualitative growth of greater richness, depth, and connection. Products will be designed with increasing efficiency (doing ever more with ever less) while simultaneously increasing their beauty, strength, and ecological integrity.

Voluntary simplicity need not imply a life of poverty, deficiency, and deprivation when living can be transformed through intelligent design into an elegant simplicity.[240] The level of satisfaction and beauty in living can be increased while lowering the quantity of resources consumed and the amount of pollution produced.

If we regard aliveness as our greatest wealth, then it is only natural for us to choose ways of living that afford greater time and opportunity to develop the areas of our lives where we feel most alive—in nurturing relationships, caring communities, time in nature, creative expression, and service to others. In seeing the Universe as alive, we naturally shift our priorities from an ego economy oriented toward consuming dead things to an aliveness economy oriented toward fostering the experience of aliveness.

An aliveness economy seeks to touch life more lightly while generating an abundance of meaning and satisfaction. Theologian Matthew Fox has written, "Luxury living is not what living is about. *Living* is what living is about! But living takes discipline and letting go and doing with less in a culture that is overdeveloped. It takes a commitment to challenge and adventure, to sacrifice and passion."[241]

In more affluent societies, consumerism is increasingly regarded as a less rewarding life pursuit and, instead, new sources of well-being are increasingly valued.[242] A major study in the US by Pew Research illustrates the growing importance of direct experience over material consumption. When asked what brings the most meaning to their lives, people replied: "spending time with family" (69%), "being outdoors" (47%), "spending time with friends" (47%), "caring for pets" (45%), "religious faith" (36%), and "job or career" (34%).

Further evidence that wealthier nations are ready to trade reduced levels of material consumption for higher levels of experiential riches is found in a study reported in the *Wall Street Journal*:

> "People think that experiences are only going to provide temporary happiness, but they actually provide both more happiness and more lasting value [than material consumption]. Experiences tend to meet more of our underlying psychological needs. They're often shared with other people, giving us a greater sense of connection, and they form a bigger part of our sense of identity."[243]

A shift toward "postmaterialist" values is also found in the highly regarded *World Values Survey* which concluded that, over a period of roughly three decades (1981–2007), a "postmodern shift" in values has been occurring in a cluster of a dozen or so nations—primarily in the United States, Canada, and Northern Europe. In these societies, emphasis is shifting from economic achievement to post-materialist values that emphasize individual self-expression, subjective well-being, and quality of life.[244]

Although simplicity has a long history, we are now entering radically changing times—ecological, social, economic, and psycho-spiritual—and we should expect the worldly expressions of simplicity to evolve and grow in response. Simplicity is not simple. A wide diversity of expressions portrays the simple life and the most useful way of describing this approach to living is with the metaphor of a garden.

To suggest the richness of simplicity, here are ten different flowerings of expression that I see growing in the "garden of simplicity." Although there is overlap among them, each expression of simplicity seems sufficiently distinct to warrant a separate category. (So there would be no favoritism, they are placed in alphabetical order based on the brief name associated with each.)

1. Artistic Simplicity: Simplicity means the way we live our lives represents a work of unfolding artistry. Leonardo da Vinci said, "Simplicity is the ultimate sophistication." Gandhi said, "My life is my message." Frederic Chopin said, "Simplicity is the final achievement. . . the crowning reward of art." In this spirit, an artistic simplicity is an understated, organic aesthetic that contrasts with the excess of consumerist lifestyles. Drawing from influences ranging from Zen to the Quakers, simplicity is a path of beauty that celebrates natural materials and clean, functional expressions.

2. Choiceful Simplicity: Simplicity means taking charge of lives that are too busy, too stressed, and too fragmented. Simplicity means choosing our unique path through life consciously, deliberately, and of our own accord. It means to live whole—to not live divided against ourselves. This path emphasizes the challenges of freedom over the comfort of consumerism. A conscious simplicity means staying focused, diving deep, and not being distracted by consumer culture. It means consciously organizing our lives so we give our "true gifts" to the world—which is to give the essence of ourselves. As Ralph Waldo Emerson said, "The only true gift is a portion of yourself."[245]

3. Compassionate Simplicity: Simplicity means to feel such a strong sense of kinship with others that, as Gandhi said, we "choose to live simply so that others may simply live." A compassionate simplicity means feeling a bond with the community of life and being drawn toward a path of reconciliation—especially with other species and future generations. A compassionate simplicity is a path of cooperation and fairness that seeks a future of mutually assured development for all.

4. Ecological Simplicity: Simplicity means to choose ways of living that touch the Earth more lightly and that reduce our ecological impact. This life-path remembers our deep roots in the natural world. It encourages us to connect with nature, the seasons, and the cosmos. A natural simplicity feels a deep reverence for the community of life on Earth and accepts that the non-human realms of plants and animals have their dignity and rights as well the human realm. Albert Schweitzer wrote, "From naïve simplicity we arrive at more profound simplicity."

5. Economic Simplicity: Simplicity means a choice for conscious consumerism and a sharing economy. Economic simplicity recognizes we are managing our relationship with our home—the Earth—and developing fitting forms of "right livelihood." It also recognizes the deep transformation in economic activity needed to live sustainably by redesigning products and services of all kinds—from housing and energy systems to food and transportation systems.

6. Family Simplicity: Simplicity means giving priority to the lives of our children and family and not to get sidetracked by our consumer society. A growing number of parents are opting out of consumerist lifestyles and seeking ways to bring life-enhancing values and experiences into the lives of their children and family.

7. Frugal Simplicity: Simplicity means that, by cutting back on spending that is not truly serving our lives and by practicing skillful management of our personal finances, we can achieve greater financial independence. Frugality and careful financial management bring increased financial freedom and the opportunity to more consciously choose our path through life. Living with less also decreases the impact of our consumption on the Earth and frees resources for others.

8. Political Simplicity: Simplicity means organizing our collective lives in ways that enable us to live more lightly and sustainably on the Earth which, in turn, involves changes in nearly every area of public life—zoning, education, transportation, and energy systems. All of these involve political choices. The politics of simplicity are also media politics as the mass media is the primary vehicle promoting the mass consciousness of consumerism.

9. Soulful Simplicity: Simplicity means to approach life as a meditation and to cultivate our experience of intimate connection with all that exists. A spiritual presence infuses the world and, by living simply, we can more directly awaken to the living universe that surrounds and sustains us, moment by moment. Soulful simplicity is more concerned with consciously tasting life in its unadorned richness than with a particular standard or manner of material living. In cultivating a soulful connection with life, we tend to look beyond surface appearances and bring our interior aliveness into relationships of all kinds.

10. Uncluttered Simplicity: An uncluttered simplicity means cutting back on trivial distractions, both material and non-material, and focusing on the essentials—whatever those may be for each of our unique lives. As Thoreau said, "Our life is frittered away by detail. . . Simplify, simplify." Or, as Plato wrote, "In order to seek one's own direction, one must simplify the mechanics of ordinary, everyday life."

As these approaches illustrate, the growing culture of simplicity contains a flourishing garden of expressions whose great diversity—and intertwined unity—are creating a resilient and hardy ecology of learning about how to live more sustainable and purposeful lives. As with other ecosystems, it is the diversity of expressions that fosters flexibility, adaptability, and resilience. Because there are so many pathways of great relevance into the garden of simplicity, this way of life has enormous potential to grow—particularly if it is nurtured and cultivated in the mass media as a legitimate, creative, and promising path for a future beyond materialism and consumerism.

14. What Can We Do Once We Know?

We have explored seven evolutionary choices at the foundation of a future of Great Transition: *aliveness, consciousness, maturity, reconciliation, communication, community, and simplicity*. Taken together, these choices can take us along a pathway of healing and wholeness, enabling us to establish ourselves as an enduring species-civilization that can evolve into the deep future. These are not easy choices, but they are authentic choices that can actualize humanity's great transition. Once we consciously understand these choices, we also recognize:

> *To not choose the Earth as our home*
> *is a profound choice to accept a pathway ahead*
> *that leads to either chaos and collapse*
> *or to authoritarianism and constraint.*

Although the times ahead will be extremely demanding, the possibility of a promising future is real and close at hand if we make foundational choices like those described in the preceding section.

At a more personal level, what can we do once we know the perils and the promise of the pathways before us? The challenges before us are so immense that we can quickly feel overwhelmed and discouraged. What choices can we make at a personal level that are meaningful and approachable? Here are some suggestions:

1. Awaken to aliveness—We are learning to live in our living universe. The most transformative action we can take is to develop our experience and appreciation of living in this great aliveness. Take walks in nature and be prepared to be surprised by the beauty and aliveness you see. Choose activities that bring you alive: dancing, playing, making music, nurturing relationships, making art, and spending time with animals. Create a small altar of gratitude with a flower, stone, leaf, feather or other memento of aliveness. Do daily affirmations or prayers for plants, animals, places, cultures, and other aspects of the living world. Become a role model of gratitude and aliveness for children.

2. Cultivate your "true gifts"—We each have "near gifts" and "true gifts." [246] Near gifts are activities we are relatively skilled at doing. We often make a living with our near gifts. True gifts express our natural talents and skills—activities where we naturally shine. True gifts draw out our inborn, soulful, natural talents and our intuitive skills. Practicing our true gifts is an exercise in becoming more fully alive and connected with the world.

3. Be deeply informed—Get to know your local ecosystem. Learn the trees, flowers, birds and animals that abound locally. Recognize the foods grown locally that you consume regularly. Explore and experience nature when you take a walk. Recognize examples of climate change and species extinction (such as disappearing insects and changing seasons). Look *wide* (consider other trends in addition to climate change), look *deep* (include both outer and inner dimensions of life), and look *long* (reflect on what life may be like five or more decades from now).

4. Protect and restore nature—After taking time to appreciate the simple gifts of nature (plants, animals, water, land, and air), take small actions to help restore these miracles of life. Be curious and learn how you can protect and restore the natural world around you. Because nature cannot advocate for itself, become a voice for nature and its preservation and restoration.

5. Grieve losses—Create an altar in your home to acknowledge, with images and objects, what we are losing (trees, flowers, animals, seasons, places, etc.). Organize a simple grieving ritual with others and have each person share something they are mourning and see being lost or forgotten—speak deeply, sing songs, read poetry, and share art.

6. Practice reconciliation—Acknowledge both privileges and inequities. In many developed countries, a person earning roughly a teacher's salary is living within the top 1% of the world's income. Recognizing your advantages, explore what this means with a group of trusted friends or peers. Identify your sphere of influence, the resources you have, and voice your concerns. Beyond economic privilege, bring curiosity and compassion to divisions of gender, race, wealth, religion, and sexual orientation.

7. Choose simplicity—Buy fewer things, give more away, eat lower on the food chain, travel less by plane, reduce or change your commute, and share your resources with others in need. Develop non-material riches that enhance your life. Cultivate meaningful friendships, share simple meals together, take more walks in nature, make music, do art, learn to dance, develop your inner life (meditation, prayer, and yoga).

8. Organize a study group—Step back and look at our world in a time of unprecedented transition. How would a more mature species behave? How would reconciliation show up more fully in your world? How could we use our powerful tools of communication to advance our maturation and reconciliation as a species? How could we evolve our physical communities for a world in transition? Do you see changes in the seasons, plants and animals? Try to avoid jumping into problem-solving or blaming and leave plenty of room for feelings to be expressed.

9. Support Others—Encourage and assist individuals and communities directly impacted by climate change through volunteering, donations, and making their situation more visible. Make your life a statement of care by acting to protect the local ecology. Volunteer for service organizations—a local foodbank, homeless shelter, or ecology group. Support young people who are living into this time of great transition. Volunteer, listen, and encourage their leadership.

10. Cultivate communication—Become a voice for the Earth and humanity's future. Contribute to newsletters, blogs, articles, videos, podcasts, and radio to bring your voice and views to our endangered future. Help awaken our social imagination to the choices we have for maturation, reconciliation, community, and simplicity. Support "Community Voice" initiatives as well as those for an "Earth Voice."

11. Become a compassionate activist—Join with others working for deep transformation. Search the internet to find organizations that fit your interests; for example, the Sunrise Movement, the Extinction Rebellion, School Strike for Climate, Youth Climate Movement, Deep Adaptation, 350.org, the Union of Concerned Scientists, Greenpeace, Natural Resources Defense Council, the Yale Program on Climate Change, and many more—including the Choosing Earth Project associated with this book: www.ChoosingEarth.org. Whether local or global, find a community that supports you in bringing your true gifts into the world at this critical time. Give of your time, your love, your talents, and your resources. If there were ever a time to step forward for the future, it is now!

When we look beyond our personal lives we can ask, "What can we do to bring our collective lives into alignment with the regenerative needs of the Earth?" In asking this key question, it is important to recognize that consumer societies are currently asking entirely different questions: Corporations ask what will produce ever-greater profits. Media giants ask what will generate the most advertising revenue. Governments ask what will promote economic growth. Educational institutions ask what skills students need to get jobs and serve a growing economy. One by one, we see that our major institutions are not oriented toward serving life and the regenerative needs of the Earth; instead, they are serving the short-term wants of consumer society.

Overall, consumer societies are unprepared for the great transition and the task of creating a regenerative approach to the future. Therefore, when we ask, "What can we do collectively?" a key answer is that *we can hold major institutions (business, media, government, and education) publicly accountable for recognizing and responding to the critical challenges facing the Earth and humanity's future.* Accountability is challenging because we all are embedded within these institutions—which means we are holding ourselves accountable as well.

. . .

Summarizing the journey we have taken in this short book: We have expanded our capacity for social imagination by looking a half-century into the future and explored unprecedented challenges and opportunities that await of us. We recognize we have a large measure of choice in the world that unfolds from here.

Choices that can support us moving
through our time of Great Transition include:
Turning toward the true wealth of aliveness
Awakening to a more conscious species-civilization
Maturing into our early adulthood
Working for broad and deep reconciliation
Developing full-hearted communication
Cultivating caring communities
Choosing lifeways of elegant simplicity

With these soulful choices, we can grow into a new experience of reality, human identity, and evolutionary journey enabling us to establish ourselves as an enduring species-civilization that can evolve into the deep future.

APPENDIX:

From "Community Voice"
to "Earth Voice"

Communication is the lifeblood of democracy, civilization, and creative evolution. Without a rich flow of communication in these times of great trauma and transition, we will perish.

The climate crisis is a communication crisis.

We cannot "fix" the climate crisis and return to the past. We must move into a world beyond crisis and make peace with the Earth's natural ecology—and this requires an extraordinary increase in the scope and depth of human communication and understanding. We require both an *informed* public and an *engaged* public. If we are informed but not engaged, the result can be frustration, alienation, and anger. If we are engaged but not informed, the result can be confusion and fruitless action.

We can achieve a new scope and depth of communication by more consciously using the tools already present in our lives. Consider the two, main tools of communication we now use daily: television and the internet.

- Television has great breadth of reach but is generally shallow.
- The internet has great depth of reach but is generally narrow.

In isolation from one another, the result is communication that tends to be shallow and narrow. However, if the power of each is combined, we can awaken communication that is deep and wide! These are not competing technologies but complementary and highly synergistic. The tools for a revolution in communication surround us if we will make conscious use of them.

The internet and television can come together to create a new level of public dialogue in response to the climate crisis. We can build on hundreds of years of experience in the US with the "New England Town Meetings" where the residents of a town vote on issues of common concern. In the modern era, we can consider an entire metropolitan area (San Francisco, Boston, Miami, etc.) as a "town" and the residents of that area eligible to "vote" and offer their advisory views on key concerns such as the climate crisis.

Until we had access to broadcast television and the internet, a "town meeting" of such scale would have been unthinkable. Now this kind of "Electronic Town Meeting," or ETM, is entirely practical.[247] A metropolitan scale Electronic Town Meeting is not

162

a fantasy—the workability of this approach was demonstrated decades ago in the San Francisco Bay Area. In 1987 a prime-time ETM on broadcast television was developed through the cooperative efforts of a non-partisan media organization I co-founded. Living in the Bay Area, our organization was called "Bay Voice."

To build our community voice organization, we brought together a diverse coalition of citizen groups. These groups ranged from different ethnic groups, business and labor, women and men, environmentalists and developers, and so on. This coalition was broad enough to genuinely reflect the diverse views and interests of the Bay Area community. The community coalition then worked with the local ABC-TV station to produce the pilot ETM.

Prior to the ETM, we worked with two major universities (Stanford and UC Berkeley) to identify a scientific, random sample of citizens who could participate by giving feedback from their homes. Those who agreed were sent a list of phone numbers that corresponded to various options they could dial in (this experiment was conducted more than a decade before the internet became widely used).

The pilot ETM began with an informative, mini-documentary to place our issue in context, and then moved to in-studio dialogue with experts and a diverse studio audience. As key questions arose in the studio discussion, they were presented to the scientific sample viewing the ETM. They dialed-in votes which were then displayed to participants in the studio and viewers at home. Six votes were easily taken during the prime-time, hour-long ETM that was viewed by more than 300,000 persons in the Bay Area. With six votes, the overall views and attitudes of the Bay Area public were clearly established. (See the first 3½ minutes of this video clip.)[248]

Decades later, in the internet era, the feedback process would be much easier to obtain. The success of this pilot just begins to demonstrate the potential for achieving a dramatic increase in the scope and depth of metropolitan-scale dialogue and consensus building. This practical capacity is now relevant for meeting the climate crisis. For example, unless actions are taken quickly, the San Francisco Bay Area will experience countless billions of dollars in flooding damages within the next few decades. A similar situation faces Miami, Boston, and a number of other vulnerable seacoast cities in the US and around the world.

Mobilizing hundreds of thousands or even millions of citizens in collective conversation with meaningful feedback requires a new scope and depth of citizen communication—and this is where non-partisan Electronic Town Meetings can be immensely valuable. By combining the broad reach of television with the penetrating depth of the internet, we already possess the technologies necessary for a revolution in civic communication. We are needlessly diminishing the power of democracy by not using these powerful tools to serve our needs as citizens. This is not an idealistic dream. In the US, metropolitan-scale Electronic Town Meetings are a direct expression of the legal rights of citizens.[249]

It is not the purpose of ETMs to have the public become directly involved in complex policy decisions; instead it is to enable citizens to express their overall views that can guide policymaking. For example, as the climate challenge grows, it is important for elected representatives to know public sentiments regarding the use of solar power, wind generation, conservation measures and other approaches for renewable energy. Assuming public feedback is advisory, ETMs respect the responsibility of elected leaders to make decisions and the responsibility of citizens to communicate effectively with those who govern.

Around the world, it is the major metropolitan areas that are the natural scale for organizing a new level of citizen dialogue. If individual communities were to form independent, non-partisan "Community Voice" organizations to launch Electronic Town Meetings, it could revolutionize the conversation of democracies within a matter of months. The leadership of one community could inspire other communities to create their own Community Voice organization and an entirely new layer of sustained and meaningful dialogue could quickly sweep across countries and around the Earth. Citizens could voice their views, propose and debate solutions, and help break through the gridlock at local, national, and international levels.

The human community has entered uncharted territory. We have never before had to come together in such elevated purpose as communities, countries, and world. A perfect storm of global crises is growing in intensity and challenging us to make dramatic changes in our manner of living. We can make this transition swiftly by building a new level of civic understanding and dialogue.

Let's go to work: Fundamentally, major metropolitan areas need to create an *independent, strictly non-partisan* "community voice" organization that has a diversity of representatives drawn from the community and that genuinely reflects the people of that area. The key responsibilities are simple and straightforward: 1) conduct research to determine which are the priority concerns of the community, and 2) work with television stations, internet providers, and survey researchers to develop the scientific sample for the ETMs, with an appropriate format.

The "community voice" organization does not promote or advocate any outcome; rather, its goal is to support community learning, dialogue, and consensus building—and then let the chips fall where they may.

In getting work underway, perhaps no factor will have a greater impact on the design, character, and implementation of ETMs than who sponsors them. Consider three major possibilities.

- First, if ETMs are sponsored by commercial TV stations, they will be designed to sell advertising and entertain audiences—not to inform citizens and involve the public in choosing its future.

- Second, if ETMs are sponsored by a local, state or national government, there will be a natural tendency to use ETMs as a public relations tool rather than as an authentic forum for open dialogue by the community.

- Third, if ETMs are sponsored by an issue-oriented organization or by an institution representing a particular ethnic, racial or gender group, then there will be a tendency to focus on the concerns of this group.

A critical conclusion emerges: an independent, non-partisan social organization is needed to act on behalf of all citizens as the sponsor of Electronic Town Meetings.

Looking beyond metropolitan-scale ETMs, once three or more community voice organizations are established, it would be practical for them to join together to create regional ETMs; for example, seacoast cities joining in common cause to respond to sea level rise.

Looking beyond regional Electronic Town Meetings, both national and global ETMs are already technologically practical. An *Earth Voice* movement is feasible as, in recent years, between 3 and 4 billion people watch the Olympics on television globally.[250] Because this number of people could be connected through the internet with mobile phones as feedback devices, *it is already technologically practical for roughly half of the world's population to engage in a global Electronic Town Meeting!*

The barrier to an Earth Voice movement is not technology; rather, the barrier is mobilizing our collective social will to come together and provide a common voice for our pathway ahead. Earth Voice dialogues could emerge rapidly from the combined initiatives of local Community Voice organizations that work together to catalyze global dialogues. The collective intelligence of our global nervous system could rapidly take on a life of its own in a process that transcends artificial, nation-state boundaries.

Like metropolitan-scale ETMs, an Earth Voice movement recognizes that its legitimacy and power depend on honoring a few, key principles:

- No advocacy by the independent, non-partisan Earth Voice organization

- Trans-partisan presentation of diverse world views in the ETM

- One person, one vote with accurate representation assured, perhaps by using block-chain technologies and validated with scientific sampling

- No attachment to outcome by the Earth Voice organization

In conclusion, the next great superpower will be the billions of ordinary citizens who encircle the Earth and who call, with a collective voice, for unprecedented cooperation and creative action to care for our endangered Earth.

Acknowledgements

Research, writing and outreach for *Choosing Earth* has been supported by the courageous and generous funding of the Roger and Brenda Gibson Family Foundation. Roger and Brenda have been key allies and soulful friends in this intensely demanding work. I would not have been able to complete this book—the culmination of a lifetime of research, writing, and learning—without their support, friendship, and trust in me. They have not only supported this book but also the larger project and learning resources that go with it. I am profoundly grateful for their partnership in helping birth this work and bring it into the world.

My appreciation as well to Fred and Elaine LeDrew for their contributions to this pioneering work. Their modest yearly donation to me is large in its message of support and love.

This book is an integral part of the larger "Choosing Earth Project" that is described in our website: www.ChoosingEarth.org. While I have been writing this book over the last year and a half, my partner and wife, Coleen LeDrew Elgin, led the development of the overall project; including pilot workshops, a learning community, and a package of education materials—videos, infographics, curriculum, timeline, PowerPoint presentation, study guide and website. She is also developing a brief, video news program on these themes. This project could not have taken root without Coleen's invaluable and skillful efforts for which I am tremendously grateful.

Sandy Wiggins has been a key member of our team from its inception. Sandy's participation and friendship have been vital as he helped create the curriculum for our basic workshops and provided feedback and suggestions for educational materials. He is a key collaborator on the East coast where he leads workshops and presentations.

A deep bow of gratitude for the thoughtful feedback and discerning suggestions the following persons offered to this book: Coleen LeDrew Elgin, Laura Loescher, Roger Gibson, Brenda Gibson, David Christel, Eden Trenor, Sandy Wiggins, Ben Elgin, Scott Elrod, Marga Laube, and Liz Moyer.

My great appreciation to Birgit Wick for bringing her artistry, skills and patience to the design and layout of this book as well as to the cover design and other materials in this project.

About the Author

Duane Elgin is an internationally recognized author, speaker, educator, and citizen-voice activist. He has an MBA from the Wharton Business School, and an MA in economic history from the University of Pennsylvania. He received Japan's Goi Peace Award in Tokyo in 2006 in recognition of his contribution to a global "vision, consciousness, and lifestyle" that fosters a "more sustainable and spiritual culture."

His books include: ***Voluntary Simplicity:*** *Toward a Way of Life that is Outwardly Simple, Inwardly Rich*; ***Awakening Earth:*** *Exploring the Evolution of Human Culture and Consciousness*; ***Promise Ahead:*** *A Vision of Hope and Action for Humanity's Future*; and ***The Living Universe:*** *Where Are We? Who Are We? Where Are We Going?* With Joseph Campbell and other scholars, he co-authored the book ***Changing Images of Man***. In addition, he has contributed chapters to more than two dozen books and published more than a hundred major articles.

In the 1970s, Duane worked as a senior staff member of the Presidential Commission on the American Future. He then worked as a senior social scientist with the think tank SRI International where he co-authored numerous studies of the long-range future; for example, *Anticipating Future National and Global Problems* (for the National Science Foundation), *Alternative Futures for Environmental Policy* (for the EPA), and *Limits to the Management of Large, Complex Systems* (for the President's Science Advisor).

As a media activist since 1981, he has co-founded three non-profit and trans-partisan organizations working for media accountability, citizen empowerment, and "Electronic Town Meetings."

As a speaker, Duane has given over three hundred keynote presentations and workshops with audiences ranging from business executives and civic groups to churches and college students.

In 2001 he was awarded an honorary PhD for work in "ecological and spiritual transformation" from the California Institute of Integral Studies (CIIS) in San Francisco.

See websites: www.DuaneElgin.com and www.ChoosingEarth.org

CHOOSING EARTH | About the Author   171

Endnotes

1   Gus Speth, quoted in the Canadian Association of the Club of Rome, March 27, 2016. https://canadiancor.com/scientists-dont-know/

2   Timothy M. Lenton, et. al., "Climate tipping point—too risky to bet against," *Nature*, November 27, 2019. https://www.nature.com/articles/d41586-019-03595-0 Also:

    "Global Risk Report, 2019" (14th Edition), *World Economic Forum*, 2019. https://www.mmc.com/insights/publications/2019/jan/global-risks-report-2019.html Julia Conley, "After Hottest Decade Since Records Began, WMO Warns World May Face 5°C Rise by Century's End," Common Dreams, December 3, 2019. "As the decade comes to a close, the world's top climate scientists warned Tuesday that policymakers' continued failure to curb the warming of the planet could lead to a global temperature increase of 5° Celsius by the end of the century and put the world 'nowhere near on track' to avoid the worst impacts of the climate crisis" https://www.commondreams.org/news/2019/12/03/after-hottest-decade-records-began-wmo-warns-world-may-face-5degc-rise-centurys-end

3   In 1978 I began writing about the decade of the 2020s, viewing it as a pivotal time for the human species. My initial descriptions of this decade (and beyond) were developed in a scenario of the next 150 years that was the basis for a script for a min-series to be produced by ABC Television. The mini-series was titled "The People of the Earth" and, although scripted, did not go into production. I continued writing about the 2020s as a pivotal decade and this is described in my books, *Awakening Earth: Exploring the Evolution of Human Culture and Consciousness*, Morrow, 1993, p. 253 https://duaneelgin.com/wp-content/uploads/2016/03/AWAKENING-EARTH-e-book-2.0.pdf; *Promise Ahead: A Vision of Hope and Action for Humanity's Future*, Harper/Quill, 2000, p.37; *The Living Universe: Where Are We? Who Are We? Where Are We Going?* Berrett-Koehler, 2009, p. 139; *Voluntary Simplicity: Toward a Way of Life that is Outwardly Simple, Inwardly Rich*, Harper, 1993 (revised edition), p. 175. Also:

    "Collective Consciousness and Cultural Healing," *Fetzer Institute*, October 1997, p. 18. Available on my website: https://duaneelgin.com/wp-content/uploads/2010/11/collective_consciousness.pdf

    Duane Elgin, "The 2020 Challenge: Evolutionary Bounce or Crash?" Union Theological Seminary, New York, 1998-1999. https://ff6363be-bf10-4e58-a513-667403971339.filesusr.com/ugd/c7bea6_5cc4096f90264ef5a9830dbf19d32fae.pdf

4   *The Commission on Population Growth and the American Future*, US National Library of Medicine, 1973. https://www.ncbi.nlm.nih.gov/pubmed/12257905 Also: https://www.sourcewatch.org/index.php/Rockefeller_Commission_on_Population_Growth_and_the_American_Future

5   Joseph Campbell, et al., *Changing Images of Man*, Stanford Research Institute, Center for the Study of Social Policy, May 1974, Contract URH (489)-2150, and later published as a book by the same name, Pergamon Press, 1984. https://archive.org/details/ChangingImagesOfMan

Also: https://www.scribd.com/document/34343238/Changing-Images-of-Man-SRI-International

6   Duane Elgin, co-author, "Assessment of Future National and International Problem Areas," National Science Foundation, SRI International Report 4676, Contract: NSF/STP76-02573, February 1977.

7   Duane Elgin, "Limits to the Management of Large, Complex Systems." This report was written as part of the larger study: Op. Cit., "Assessment of Future National and International Problem Areas," National Science Foundation, Contract: NSF/STP76-02573, February 1977. https://duaneelgin.com/wp-content/uploads/2014/11/Limits-to-Large-Complex-Systems.pdf

8   Duane Elgin, David MacMichael, Peter Schwartz, "Alternative Futures for Environmental Policy Planning: 1975–2000, Environmental Protection Agency, Washington, DC, Contract 68-01-2698, October 1975. https://www.worldcat.org/title/alternative-futures-for-environmental-policy-planning-1975-2000/oclc/01968862

9   Duane Elgin and Arnold Mitchell, "Voluntary Simplicity," SRI International, Business Intelligence Program, 1976. This was published as a revised article in *The Co-Evolution Quarterly* in 1977. https://duaneelgin.com/wp-content/uploads/2010/11/voluntary_simplicity.pdf Also see my book: *Voluntary Simplicity: Toward a Way of Life that is Outwardly Simple, Inwardly Rich*, Harper, 2010 (second revised edition).

10   For example, see: Duane Elgin, "Revitalizing Democracy Through Electronic Town Meetings" *Spectrum: The Journal of State Governments*, Spring, 1993. https://duaneelgin.com/wp-content/uploads/1993/12/ETMs-Spectrum-Journal.pdf

11   See, for example, the following sources regarding global transition:

Joseph Tainter, *The Collapse of Complex Societies*, Cambridge University Press, 1988.

Donnela Meadows, et. al., *Beyond the Limits*, Chelsea Green, 1992.

Duane Elgin, *Awakening Earth*, William Morrow, 1993.

Jared Diamond, *Collapse: How Societies Choose to Fail or Succeed*, Penguin Group, 2005.

James Howard Kuntzler, *The Long Emergency*, Grover/Atlantic, 2005.

Dmitry Orlov, *Reinventing Collapse*, New Society Publishers, 2008.

Carolyn Baker, *Sacred Demise: Walking the Spiritual Path of Industrial Civilization's Collapse*, iUniverse 2009

Jem Bendell, *Deep Adaptation: A Map for Navigating Climate Tragedy*, IFLAS Occasional Paper 2, December 2018. https://www.lifeworth.com/deepadaptation.pdf

Paul Gilding, *The Great Disruption*, Bloomsbury, 2012.

Ervin Laszlo, *World Shift*, Inner Traditions, 2009.

Richard Heinberg, *Peak Everything*, New Society Publishers, 2010.

John Michael Greer, *Dark Age America*, New Society Publishers, 2016.

Paul Raskin, *The Great Transition Initiative*, https://greattransition.org/

Luke Kemp, "Are we on the road to civilisation collapse?" *BBC Future*, February 18, 2019. https://www.bbc.com/future/article/20190218-are-we-on-the-road-to-civilisation-collapse

Luke Kemp, "The Lifespan of Ancient Civilizations," *BBC Future*, February 19, 2019. https://www.bbc.com/future/article/20190218-the-lifespans-of-ancient-civilisations-compared

Rachel Nuwer, "How Western civilisation could collapse," *BBC Future*, April 17, 2017. https://www.bbc.com/future/article/20170418-how-western-civilisation-could-collapse

12 Betsy MacGregor, *In Awe of Being Human: A Doctor's Stories from the Edge of Life and Death*, Abiding Nowhere Press, 2013. https://www.amazon.com/Awe-Being-Human-Doctors-Stories/dp/0985496770

13 Jacques Verduin, Founder of the GRIP Program (Guiding Rage Into Power) GRIP is a deeply transformative program for life-sentenced prisoners. https://insight-out.org/ Amor fati (lit. "love of fate") is a Latin phrase that may be used to describe an attitude in which one sees everything that happens in one's life, including suffering and loss, as good or, at the very least, necessary." https://en.wikipedia.org/wiki/Amor_fati

14 The evolution from an "Earth Goddess" perspective to a "Sky God" perspective to the rise of the "Cosmic Goddess" is explored in my book, *Awakening Earth*, Op. Cit, 1993. https://duaneelgin.com/wp-content/uploads/2016/03/AWAKENING-EARTH-e-book-2.0.pdf

15 Regarding the soul of the universe from the perspective of a feminine archetype has been developed by Anne Baring. See her magnificent book, *The Dream of the Cosmos*, Archive Publishing, 2013. Free download at: https://all-med.net/pdf/the-dream-of-the-cosmos/

16 "World Scientists' Warning to Humanity," *Union of Concerned Scientists*, 1992 onwards. https://www.ucsusa.org/resources/1992-world-scientists-warning-humanity

17 Ibid.

18 Duane Elgin, *Awakening Earth* (1993), Op. Cit.; *Voluntary Simplicity* (second edition, 1993); *Promise Ahead* (2000), Op. Cit.; and *The Living Universe* (2009), Op. Cit.

19 World Scientists,' "Warning to Humanity: A Second Notice;" https://academic.oup.com/bioscience/article/67/12/1026/4605229

20 William J Ripple, et al., "World Scientists' Warning of a Climate Emergency," *BioScience*, November 5, 2019 biz088, https://doi.org/10.1093/biosci/biz088 Also:

"Emissions Gap Report," United Nations Environment Program, November 2019. https://www.unenvironment.org/resources/emissions-gap-report-2019

"In bleak report, U.N. says drastic action is only way to avoid worst effects of climate change," Washington Post, November 26, 2019. https://www.washingtonpost.com/climate-environment/2019/11/26/bleak-report-un-says-drastic-action-is-only-way-avoid-worst-impacts-climate-change/

Somini Sengupta, "World Powers Vowed to Cut Greenhouse Gases. They're Still Rising Perilously," https://governorswindenergycoalition.org/world-powers-vowed-to-cut-greenhouse-gases-theyre-still-rising-perilously/

21 Rachel Carson's commencement address at Scripps College in California, June 1962 titled, "Of Man and the Stream of Time" http://www.scrippscollege.edu/news/features/in-today-already-walks-tomorrow Also see: https://www.brainpickings.org/2019/04/12/rachel-carson-scripps-college-commencement/

22 "Workers Flee and Thieves Loot Venezuela's Reeling Oil Giant," *The New York Times*, June 14, 2018. https://www.nytimes.com/2018/06/14/world/americas/venezuela-oil-economy.html

23 See, for example: Future of Life Institute, https://futureoflife.org/background/existential-risk/

24 "The Beginning of the End," The editors of the journal, *New Scientist*, October 13, 2018. https://www.newscientist.com/article/mg24031992-900-weve-missed-many-chances-to-curb-global-warming-this-may-be-our-last/

25 Owen Gaffney, "Quit Carbon, and Quick," *New Scientist*, January 5, 2019. https://www.sciencedirect.com/science/article/abs/pii/S0262407919300181

26 "Life Beyond Growth: The history and possible future of alternatives to GDP-measured Growth-as-Usual," Alan AtKisson, Stockholm, Sweden, January 31, 2012. http://www.oecd.org/site/worldforumindia/ATKISSON.pdf Even these estimates may underestimate the cost of climate change. Also:

Naomi Oreskes and Nicholas Stern, "Climate Change Will Cost Us Even More Than We Think," *The New York Times*, Oct. 23, 2019. https://www.nytimes.com/2019/10/23/opinion/climate-change-costs.html

27 "The Human Footprint," *World Wildlife Fund*, https://www.worldwildlife.org/threats/the-human-footprint Also: *Global Footprint Network*: https://www.footprintnetwork.org/

28 Donella Meadows, et. al., *Limits to Growth*, Potomac Associates, 1972. https://en.wikipedia.org/wiki/The_Limits_to_Growth

29 Eugene Linden, "How Scientists Got Climate Change So Wrong," *The New York Times*, November 8, 2019. https://www.nytimes.com/2019/11/08/opinion/sunday/science-climate-change.html Also:

"Climate Change Speed-Up," *Atmospheric Sciences & Global Change Research Highlights*, March 2015. Increasing temperature change over next several decades will accelerate, according to new research. Earth's temperature changes are happening faster than historical levels and are starting to speed up. https://www.pnnl.gov/science/highlights/highlight.asp?id=3931

"How fast is the climate changing? It's happened within one lifetime." David Wallace-Wells, climate journalist and author of "The Uninhabitable Earth," explains: https://www.youtube.com/watch?v=RA4mIbQo52k

30 Although time-scales of events described as "abrupt" may vary dramatically, there is very disturbing evidence they can be on the timescale of years! For example: "Changes recorded in the climate of Greenland at the end of the Younger Dryas [roughly 11,800 years ago], as measured by ice-cores, imply a sudden warming of +10°C (+18°F) within a timescale of a few years." Grachev, A.M.; Severinghaus, J.P., *Qauternary Science Reviews*, March, 2005. "A revised +10±4°C magnitude of the abrupt change in Greenland temperature at the Younger Dryas termination using published GISP2 gas isotope data and air thermal diffusion constants." https://ui.adsabs.harvard.edu/abs/2005QSRv...24..513G/abstract

31 An exception is Sweden: Christian Ketels and K. Persson, "Sweden's ministry for the future: how governments should think strategically and act horizontally," Centre for Public Impact, November 29, 2018. https://www.centreforpublicimpact.org/swedens-ministry-for-the-future-how-governments-should-think-strategically-and-act-horizontally/

32 Duane Elgin, co-author of the SRI International Report 4676, "Assessment of Future National and International Problem Areas," National Science Foundation, February 1977.

33 Duane Elgin, "Limits to Complexity: Are Bureaucracies Becoming Unmanageable," *The Futurist*, December 1977. https://duaneelgin.com/wp-content/uploads/2014/11/Limits-to-Large-Complex-Systems.pdf

34 Gus Speth, Op. Cit., quoted in the Canadian Association of the Club of Rome, March 27, 2016. https://canadiancor.com/scientists-dont-know/

35 Fra Giovanni Giocondo (c.1435–1515) was a Renaissance pioneer, accomplished as an architect, engineer, antiquary, archaeologist, classical scholar, and Franciscan friar. https://gratefulness.org/resource/joy-fra-giovanni-peace/

36 John Vidal, "The Lost Decade: How We Awoke To Climate Change Only To Squander Every Chance To Act," *HuffPost*, December 30, 2019. https://www.

huffpost.com/entry/lost-decade-climate-change-action-2020_n_5df7af92e
4b0ae01a1e459d2

37 Maria Repnikova, "China's 'responsive' authoritarianism," *Washington Post*,
    November 27, 2019. https://www.washingtonpost.com/news/theworldpost/
    wp/2018/11/27/china-authoritarian/ Also:

    Paul Mozur and Aaron Krolik, "A Surveillance Net Blankets China's Cities,
    Giving Police Vast Powers," *The New York Times*, December 17, 2019.
    https://www.nytimes.com/2019/12/17/technology/china-surveillance.
    html?action=click&module=Top%20Stories&pgtype=Homepage

38 Nicholas Wright, "How Artificial Intelligence Will Reshape the Global Order,"
    Foreign Affairs, July 10, 2018. https://www.foreignaffairs.com/articles/
    world/2018-07-10/how-artificial-intelligence-will-reshape-global-order

39 Duane Elgin, *The Living Universe*, Op. Cit., p. 141–142.

40 Moira Fagan, et. al., "A look at how people around the world view climate
    change," PEW Research, April 18, 2019. https://www.pewresearch.org/
    fact-tank/2019/04/18/a-look-at-how-people-around-the-world-view-
    climate-change/

41 Ibid., 2019.

42 Another indication of the danger ahead is described in the IPCC report
    on climate change and land. See: "The world has just over a decade to get
    climate change under control, U.N. scientists say." *Washington Post*, Chris
    Mooney and Brady Dennis, Oct. 7, 2018. There is no documented historic
    precedent" for the scale of changes required, the body found. Here is an
    important response to the new IPCC report on "Climate Change and Land."
    1.5 degrees is the new 2 degrees," Jennifer Morgan, executive director of
    Greenpeace International. Specifically, the document finds that instabilities
    in Antarctica and Greenland, which could usher in sea-level rise measured
    in feet rather than inches, "could be triggered around 1.5°C to 2°C of global
    warming." Moreover, the total loss of tropical coral reefs is at stake because
    70% to 90% are expected to vanish at 1.5° Celsius, the report finds. At 2°C,
    that number grows to more than 99%. The report clearly documents that a
    warming of 1.5° Celsius would be very damaging and that 2°—which used to
    be considered a reasonable goal—could produce intolerable consequences in
    parts of the world. https://www.ipcc.ch/report/srccl/ Also:

    Updated IPCC report: "New U.N. climate report: Massive change
    already here for world's oceans and frozen regions," Chris Mooney and
    Brady Dennis, *Washington Post*, September 25, 2019. https://www.
    msn.com/en-us/weather/topstories/new-u-n-climate-report-massiv
    e-change-already-here-for-world-s-oceans-and-frozen-regions/
    ar-AAHP17O?li=AAaeUIW&%253Bocid=1PRCDEFE

    "Special Report on the Ocean and Cryosphere in a Changing Climate,"
    Intergovernmental Panel on Climate Change, https://report.ipcc.ch/srocc/
    pdf/SROCC_SPM_Approved.pdf To download, go to: https://www.ipcc.ch/
    srocc/download-report/

43 "Trump administration sees a 7-degree [7° Fahrenheit or 4° Celsius] rise in global temperatures by 2100," *Washington Post*, September 28, 2018. ". . . deep in a 500-page environmental impact statement, the Trump administration made a startling assumption: On its current course, the planet will warm a disastrous seven degrees (or 4° Celsius) by the end of this century. A rise of seven degrees Fahrenheit, or about four degrees Celsius, compared with preindustrial levels would be catastrophic. . . The analysis assumes the planet's fate is already sealed." https://www.washingtonpost. com/national/health-science/trump-administration-sees-a-7-degree-rise-in-global-temperatures-by-2100/2018/09/27/b9c6fada-bb45-11e8-bdc0-90f81cc58c5d_story.html This startling conclusion is supported by more recent climate models.

Also: Eric Roston, "Climate Models Are Running Red Hot, and Scientists Don't Know Why," *Bloomberg*, February 3, 2020. https://www.bloomberg. com/news/features/2020-02-03/climate-models-are-running-red-hot-and -scientists-don-t-know-why

44 An example of damage from sea level rise is the severe erosion of the world's beaches: Half of the world's beaches could disappear by the end of the century and by 2050 some coastlines could be unrecognizable from what we see today. Michalis I. Vousdoukas, et. al., "Sandy coastlines under threat of erosion," *Nature: Climate Change*, March 2, 2020. https://www.nature. com/articles/s41558-020-0697-0

45 Impacts of global warming by 2100. See: "Six graphics explaining climate change; graphic 6 on limiting damage." Source: Climate Action Tracker, data compiled by *Climate Analytics, ECOFYS, New Climate Institute and Potsdam Institute for Climate Impact Research.* https://bbc.co.uk/news/ resources/idt-5aceb360-8bc3-4741-99f0-2e4f76ca02bb

46 *IPCC Special Report on the Ocean and Cryosphere in a Changing Climate,* IPCC, September 25, 2019. https://report.ipcc.ch/srocc/pdf/ SROCC_SPM_Approved.pdf To download, go to: https://www.ipcc.ch/ srocc/download-report/ https://climatenexus.org/climate-change-news/ ipcc-oceans-ice-systems-climate-impacts/

47 "Sea levels set to keep rising for centuries even if emissions targets met," *The Guardian*, November 6, 2019. The lag time between rising global temperatures and the impact of coastal inundation means that the world will be dealing with ever-rising sea levels into the 2300s, regardless of prompt action to address the climate crisis, according to the new study. https://www. theguardian.com/environment/2019/nov/06/sea-level-rise-centuries-climate-crisis See the study "Attributing long-term sea-level rise to Paris Agreement emission pledges,": https://www.pnas.org/content/early/2019/ 10/31/1907461116 Also:

Zeke Hausfather, "Common Climate Misconceptions: Atmospheric Carbon Dioxide," *Yale Climate Connections*, December 16, 2010. This study found that, while a good portion of greenhouse gas emissions could be removed from the atmosphere in a few decades, even if emissions were somehow ceased immediately, about 10% would continue warming Earth for thousands of years. This 10% is significant, because even a small increase in atmospheric greenhouse gases can have a large impact on ice sheets and sea level if it persists over the millennia. Even more important: The biggest danger is not global warming, it is the extreme weather produced by moving past tipping points which, in turn, lead to catastrophic famine and immense civic unrest. https://www.yaleclimateconnections.org/2010/12/common-climate-misconceptions-atmospheric-carbon-dioxide/

48 "BP Statistical Review of World Energy," *British Petroleum*, (68th edition), 2019. https://www.bp.com/en/global/corporate/news-and-insights/press-releases/bp-statistical-review-of-world-energy-2019.html

49 "Hothouse Earth Fears," *New Scientist*, August 11, 2018. https://www.sciencedirect.com/journal/new-scientist/vol/239/issue/3190 "For most of the past half billion years, Earth has been much hotter than today, with no permanent ice at the poles: the hothouse Earth state. Then, roughly three million years ago, as CO_2 levels fell, temperatures began oscillating between two cooler states: ice ages with great ice sheets covering much land in the northern hemisphere and interglacial periods like the present. With CO_2 increases we might be on the brink of pushing the planet out of the present interglacial state and into the hothouse Earth state. The consequences are beyond catastrophic." Also:

McGrath, "Climate change: 'Hothouse Earth' risks even if CO2 emissions slashed," BBC, August 5, 2018. https://www.bbc.com/news/science-environment-45084144

"New Climate Risk Classification Created to Account for Potential "Existential" Threats," *Scripps Institute of Oceanography*, September 14, 2017. "A temperature increase greater than 3°C (5.4°F) could lead to what the researchers term "catastrophic" effects, and an increase greater than 5°C (9°F) could lead to "unknown" consequences which they describe as beyond catastrophic including potentially existential threats. The specter of existential threats is raised to reflect the grave risks to human health and species extinction from warming beyond 5°C, which has not been experienced for at least the past 20 million years." https://scripps.ucsd.edu/news/new-climate-risk-classification-created-account-potential-existential-threats

Will Steffen, et. al., "Trajectories of the Earth System in the Anthropocene," *PNAS: Proceedings of the National Academy of Sciences*, August 14, 2018. "We explore the risk that self-reinforcing feedbacks could push the Earth System toward a planetary threshold that, if crossed, could prevent stabilization of the climate at intermediate temperature rises and cause continued warming on a 'Hothouse Earth' pathway even as human emissions are reduced. Crossing the threshold would lead to a much higher global average temperature than any interglacial in the past 1.2 million years and to sea levels significantly higher than at any time in the Holocene." https://doi.org/10.1073/pnas.1810141115

50 "Climate Change: How Do We Know?" *NASA: Global Climate Change, Vital Signs of the Planet*, 2019. See the evidence here: https://climate.nasa.gov/ evidence/ See scientific consensus regarding climate warming here: https://climate.nasa.gov/scientific-consensus/ Also:

"Climate change: Disruption, risk and opportunity," *Science Direct* (originally published in *Global Transitions*, Volume 1, 2019, Pages 44-49). The study concludes: Climate change is disruptive because humans have adapted to a narrow range of environmental conditions. Change is particularly risky in the presence of low predictability, large scale, rapid onset and lack of reversibility. https://doi.org/10.1016/j.glt.2019.02.001

"Global Warming Science: The science is clear. Global warming is happening." *Union of Concerned Scientists*, 2019. https://www.ucsusa.org/ our-work/global-warming/science-and-impacts/global-warming-science

Op. Cit., *IPCC Special Report on Oceans and the Cryosphere*, September 25, 2019.

Bob Berwyn, "Ocean Warming Is Speeding Up, with Devastating Consequences, Study Shows," *Inside Climate News*. January 14, 2020. In 25 years, the oceans have absorbed heat equivalent to the energy of 3.6 billion Hiroshima-size atom bomb explosions, the study's lead author said. https://insideclimatenews.org/news/14012020/ocean-hea t-2019-warmest-year-argo-hurricanes-corals-marine-animals-heatwaves

Sabrina Shankman, "Dead Birds Washing Up by the Thousands Send a Warning About Climate Change," *Inside Climate News*, January 15, 2020. A new study unravels the mystery of what caused so many of these normally resilient seabirds to starve amid an ocean heat wave fueled in part by global warming. https://insideclimatenews.org/news/15012020/seabird-deat h-ocean-heat-wave-blob-pacific-alaska-common-murre

51 "Urgent health challenges for the next decade," *WHO (World Health Organization)*, January 13, 2020. https://www.who.int/news-room/ photo-story/photo-story-detail/urgent-health-challenges-fo r-the-next-decade

52 "Powerful actor, high impact bio-threats—initial report," *Wilton Park/UK*, November 9, 2018. https://www.wiltonpark.org.uk/wp-content/uploads/ WP1625-Summary-report.pdf Also:

Nafeez Ahmed, "Coronavirus, Synchronous Failure and the Global Phase-Shift," *Insurge Intelligence*, March 2, 2020. https://www.resilience. org/stories/2020-03-05/coronavirus-synchronous-failure-and-the-glob al-phase-shift/

Jennifer Zhang, "Coronavirus Response Shows the World May Not Be Ready for Climate-Induced Pandemics," *Columbia University*, February 24, 2020. https://blogs.ei.columbia.edu/2020/02/24/coronavirus-climate-induced- pandemics/

Brian Deese and Ronald Klain, "Another deadly consequence of climate change: The spread of dangerous diseases," *Washington Post*, May 30, 2017. https://www.washingtonpost.com/opinions/another-deadly-consequence-of-climate-change-the-spread-of-dangerous-diseases/2017/05/30/fd3b8504-34b1-11e7-b4ee-434b6d506b37_story.html

I appreciate the insights of Sandy Wiggins in differentiating between the challenges of responding to pandemics and those of climate change.

53 Another study concludes that already: "Two-thirds of the global population (4.0 billion people) live under conditions of severe water scarcity at least 1 month of the year." https://www.seametrics.com/blog/future-water/ Also:

Mesfin M. Mekonnen and Arjen Y. Hoekstra, "Four billion people facing severe water scarcity," *Science Advances*, February 12, 2016. https://advances.sciencemag.org/content/2/2/e1500323.full

Another study found that between 1995 and 2025 the areas affected by "severe water stress" expand and intensify, and the number of people living in these areas also grows from 2.1 to 4.0 billion people. They state: "continuing stress on water resources increases the risk that simultaneous water shortages might occur around the world and even trigger a kind of global water crisis." Mark Rosegrant, et.al., "World Water and Food to 2025: Dealing with Scarcity," *International Food policy Research Institute*, Washington, D.C. 2002. http://www.sidnlps.org.pk/available_online/water2025-tanveer.pdf

54 "The Water Crisis," *Water.org*, 2019. https://water.org/our-impact/water-crisis/

55 "World Water Development Report," 2019. https://www.unwater.org/publications/world-water-development-report-2019/ Also: https://water.org/our-impact/water-crisis/

56 The number of undernourished people in the world has been on the rise since 2015 and is back to levels seen in 2010–2011. http://www.fao.org/state-of-food-security-nutrition/en/ Also:

"The Hungry Planet: Global Food Scarcity in the 21st Century," *Wilson Center* Staff, August 16, 2011. https://www.newsecuritybeat.org/2011/08/the-hungry-planet-global-food-scarcity-in-the-21st-century/

57 Nafeez Ahmed, "West's 'Dust Bowl' Future now 'Locked In, as World Risks Imminent Food Crisis," *Insurge Intelligence*, January 6, 2020. https://www.resilience.org/stories/2020-01-06/wests-dust-bowl-future-now-locked-in-as-world-risks-imminent-food-crisis/

58 Anup Shah, "Poverty Facts and Stats," *Global Issues*, Updated January 7, 2013. http://www.globalissues.org/article/26/poverty-facts-and-stats#src1 Also:

Anup Shah, "Poverty Around The World," *Global Issues*, November 12, 2011. http://www.globalissues.org/print/article/4#WorldBanksPovertyEstimates Revised

59 Julian Cribb, "The coming famine: risks and solutions for global food security," April 18, 2010. https://www.sciencealert.com/the-coming-famine-risks-and-solutions-for-global-food-security ca5162en.pdf

60 "Our Food Systems Are in Crisis," *Scientific American*, October 15, 2019. https://blogs.scientificamerican.com/observations/our-food-systems-are-in-crisis/

61 Izabella Koziell, "Migration, Agriculture and Climate Change," *Food and Agricultural Organization of the United Nations*, 2017. http://www.fao.org/3/I8297EN/i8297en.pdf

62 See report "Nature's Dangerous Decline 'Unprecedented'; Species Extinction Rates 'Accelerating'" *Intergovernmental Science-Policy Platform on Biodiversity and Ecosystem Services (IPBES)*, May 22, 2019. https://www.ipbes.net/news/Media-Release-Global-Assessment Also: https://www.washingtonpost.com/climate-environment/2019/05/06/one-million-species-face-extinction-un-panel-says-humans-will-suffer-result/

63 "Plummeting insect numbers 'threaten collapse of nature,'" in *The Guardian*, February 10, 2019. https://www.theguardian.com/environment/2019/feb/10/plummeting-insect-numbers-threaten-collapse-of-nature A growing number of studies are sounding the alarm that insects around the world are in crisis. For example, one study in Germany found a 76% decrease in flying insects in just the past few decades. Another study of rainforests in Puerto Rico found insects had declined as much as 60-fold. Also:

Damian Carrington, "Car 'splatometer' tests reveal huge decline in number of insects," *The Guardian*, February 12, 2020. Research shows insect populations at sites in Europe has plunged by up to 80% in two decades. https://www.theguardian.com/environment/2020/feb/12/car-splatometer-tests-reveal-huge-decline-number-insects

Damian Carrington, "Insect apocalypse' poses risk to all life on Earth, conservationists warn," *The Guardian*, November 13, 2019. Report claims 400,000 insect species face extinction amid heavy use of pesticides. https://www.theguardian.com/environment/2019/nov/13/insect-apocalypse-poses-risk-to-all-life-on-earth-conservationists-warn

Dave Goulson, "Insect declines and why they matter," Commissioned by the *South West Wildlife Trusts*, 2019. ". . . recent evidence suggests that abundance of insects may have fallen by 50% or more since 1970. This is troubling, because insects are vitally important, as food, pollinators and recyclers amongst other things." https://www.somersetwildlife.org/sites/default/files/2019-11/FULL%20AFI%20REPORT%20WEB1_1.pdf https://doi.org/10.1016/j.biocon.2019.01.020

64 "Pollinators Help One-third Of The World's Food Crop Production," *Science Daily*, October 26, 2009. https://www.sciencedaily.com/releases/2006/10/061025165904.htm Bees are the primary initiators of reproduction among plants, as they transfer pollen from the male stamens to the female pistils. Also:

Ishan Daftardar, "Why Bee Extinction Would Mean the End of Humanity," *Science ABC*, July 3, 2015. https://www.scienceabc.com/nature/bee-extinction-means-end-humanity.html

65 Carl Zimmer, "Birds Are Vanishing from North America," *The New York Times*, September 19, 2019. https://www.nytimes.com/2019/09/19/science/bird-populations-america-canada.html

66 Kenneth Rosenberg, et. al., "Decline of the North American avifauna," *Science*, October 4, 2019. https://science.sciencemag.org/content/366/6461/120

67 J. Emmett Duffy, et. al., "Science study predicts collapse of all seafood fisheries by 2050," *Stanford Report*, November 2, 2006. https://news.stanford.edu/news/2006/november8/ocean-110806.html "All species of wild seafood will collapse within 50 years, according to a new study by an international team of ecologists and economists. . . Based on current global trends, the authors predicted that every species of wild-caught seafood—from tuna to sardines—will collapse by the year 2050. 'Collapse' was defined as a 90 % depletion of the species' baseline abundance." Also:

Jeff Colarossi, "Climate Change And Overfishing Are Driving The World's Oceans To The 'Brink Of Collapse,'" *Think Progress*, 2015. https://archive.thinkprogress.org/climate-change-and-overfishing-are-driving-the-worlds-oceans-to-the-brink-of-collapse-2d095e127640/ "Within a single generation, human activity has severely damaged almost every aspect of our global oceans. That's the finding of a new World Wildlife Fund study, which revealed that marine populations have declined 49% between 1970 and 2012. . . The picture is now clearer than ever: humanity is collectively mismanaging the ocean to the brink of collapse."

"Living Blue Planet Report: Species, habitats and human well-being," World Wildlife Fund, 2015. http://assets.wwf.org.uk/downloads/living_blue_planet_report_2015.pdf?_ga=1.259860873.2024073479.1442408269

Ivan Nagelkerken and Sean D. Connell, "Global alteration of ocean ecosystem functioning due to increasing human CO_2 emissions," *PNAS: Proceedings of the National Academy of Sciences*, October 27, 2015. https://doi.org/10.1073/pnas.1510856112

68 Adam Vaughan, "Humanity driving 'unprecedented' marine extinction," *The Guardian*, September 14, 2016. https://www.theguardian.com/environment/2016/sep/14/humanity-driving-unprecedented-marine-extinction The study can be found here: "Ecological selectivity of the emerging mass extinction in the oceans," *Science*, September 16, 2016. https://science.sciencemag.org/content/353/6305/1284

69 "Saving Life on Earth: a plan to halt the global extinction crisis," *Center for Biological Diversity*, January 2020. https://www.biologicaldiversity.org/programs/biodiversity/elements_of_biodiversity/extinction_crisis/pdfs/Saving-Life-On-Earth.pdf

70 Current UN World Population estimates. https://www.worldometers.info/world-population/

71　Rob Smith, "These will be the world's most populated countries by 2100," *World Economic Forum*, February 29, 2018. https://www.weforum.org/agenda/2018/02/these-will-be-the-worlds-most-populated-countries-by-2100/ Also:

Jeff Desjardins, "The world's biggest countries, as you've never seen them before," *World Economic Forum*, October 4, 2017. https://www.weforum.org/agenda/2017/10/the-worlds-biggest-countries-as-youve-never-seen-them-before

72　World Population Growth. Sources: *Population Division of the Department of Economic and Social Affairs of the United Nations Secretariat*, 2013 and World Population Prospects The 2012 Revision, New York, United Nations. Less developed regions: Africa, Asia (excluding Japan), Latin American and the Caribbean, and Oceana (excluding Australian and New Zealand). More developed regions: Europe, North America (Canada and the United States), Japan, Australia, and New Zealand. https://kids.britannica.com/students/assembly/view/171828

73　Bradshaw and Barry Brook, "A killer plague wouldn't save the planet from us," *New Scientist*, November 1, 2014. https://www.newscientist.com/article/mg22429934-100-a-killer-plague-wouldnt-save-the-planet-from-us/ An approximate carrying capacity of the Earth can be found in the article. The authors estimate a sustainable human population, given current Western consumption patterns and technologies, would be between 1 and 2 billion people. Also:

Another perspective is offered by the visionary scientist, James Lovelock, who believes the Earth's population will fall to as few as 500 million by 2100, with most of the survivors living in the far northern latitudes—Canada, Iceland, Scandinavia, the Arctic Basin. See the interview: Jeff Goodell, "Hothouse Earth Is Merely the Beginning of the End," *Rolling Stone* magazine, August 9, 2018. https://www.rollingstone.com/politics/politics-features/hothouse-earth-climate-change-709470/

4 Degrees Hotter, A Climate Action Centre Primer, February 2011. Melbourne, Australia. https://www.climatecodered.org/2011/02/4-degrees-hotter-adaptation-trap.html The study quotes Professor Kevin Anderson, director of the *Tyndall Centre for Climate Change*, who "believes only around 10% of the planet's population—around half a billion people—will survive if global temperatures rise by 4°C. He said the consequences were "terrifying." "For humanity it's a matter of life or death," he said. "We will not make all human beings extinct as a few people with the right sort of resources may put themselves in the right parts of the world and survive. But I think it's extremely unlikely that we wouldn't have mass death at 4°C." In 2009 Professor Hans Joachim Schellbhuber, director of the *Potsdam Institute*, and one of Europe's most eminent climate scientists, told his audience that at 4C, population ". . . carrying capacity estimates (are) below 1 billion people."

"Carrying capacity," Wikipedia, 2019. "Several estimates of the carrying capacity have been made with a wide range of population numbers. A 2001 UN report said that two-thirds of the estimates fall in the range of 4 billion to 16 billion with unspecified standard errors, with a median of about 10 billion.[5] More recent estimates are much lower, particularly if non-renewable resource depletion and increased consumption are considered." https://en.wikipedia.org/wiki/Carrying_capacity

"How many people can Earth actually support?" *Australian Academy of Science*, 2019. https://www.science.org.au/curious/earth-environment/how-many-people-can-earth-actually-support "If everyone on Earth lived like a middle class American, then the planet might have a carrying capacity of around 2 billion. However, if people only consumed what they actually needed, then the Earth could potentially support a much higher figure.

Marian Starkey, "What is the Carrying Capacity of Earth?" *Population Connection*, April 13, 2017. https://www.populationconnection.org/carrying-capacity-earth/ "Already, we're consuming the Earth's renewable resources at one and a half times the sustainable rate. And that's with billions of people living in poverty, consuming next to nothing. Imagine what would happen if desperately poor people were fortunate enough to live a middleclass lifestyle. And then imagine what would happen if poor people joined the middle class, AND the human population grew from today's 7.5 billion to 9, 10, or 11 billion."

Andrew D. Hwang, "The human population is 7.5 billion and counting—a mathematician counts how many humans the Earth can actually support," *Business Insider*, July 10, 2018. https://www.businessinsider.com/how-many-people-earth-can-hold-before-runs-out-resources-2018-7 According to the Worldwatch Institute, an environmental think tank, the Earth has 1.9 hectares of land per person for growing food and textiles for clothing, supplying wood and absorbing waste. The average American uses about 9.7 hectares. These data alone suggest the Earth can support at most one-fifth of the present population, 1.5 billion people, at an American standard of living. The Earth supports industrialized standards of living only because we are drawing down the "savings account" of non-renewable resources, including fertile topsoil, drinkable water, forests, fisheries and petroleum.

Natalie Wolchover, "How Many People Can Earth Support?," *Live Science*, October 11, 2011. https://www.livescience.com/16493-people-planet-earth-support.html "10 billion people is the uppermost population limit where food is concerned. Because it's extremely unlikely that everyone will agree to stop eating meat, E.O. Wilson thinks the maximum carrying capacity of the Earth based on food resources will most likely fall short of 10 billion."

Vivien Cumming, "How many people can our planet really support?," *BBC*, March 14, 2016. http://www.bbc.com/earth/story/20160311-how-many-people-can-our-planet-really-support "It is not the number of people on the planet that is the issue—but the number of consumers and the scale and nature of their consumption," says David Satterthwaite, a senior fellow at the International Institute for Environment and Development in London. He quotes Gandhi: "The world has enough for everyone's need, but not enough for everyone's greed." . . . The real concern would be if the people living in these areas decided to demand the lifestyles and consumption rates currently considered normal in high-income nations; something many would argue is only fair. . . . Only when wealthier groups are prepared to adopt low-carbon lifestyles, and to permit their governments to support such a seemingly unpopular move, will we reduce the pressure on global climate, resource and waste issues. . . . Only when wealthier groups are prepared to adopt low-carbon lifestyles, and to permit their governments to support such a seemingly unpopular move, will we reduce the pressure on global climate, resource and waste issues. . . .For the foreseeable future, Earth is our only home and we must find a way to live on it sustainably. It seems clear that that requires scaling back our consumption, in particular a transition to low-carbon lifestyles, and improving the status of women worldwide. Only when we have done these things will we really be able to estimate how many people our planet can sustainably hold.

"One Planet, How Many People? A Review of Earth's Carrying Capacity," *UNEP*, June 2102. https://na.unep.net/geas/archive/pdfs/geas_jun_12_ carrying_capacity.pdf While there is an incredible range to the estimates of Earth's carrying capacity, the greatest concentration of estimates falls between 8 and 16 billion people (3). Global population is fast approaching the low end of that range and is expected to get well into it at around 10 billion by the end of the century.

74 Ecological Footprint, https://www.footprintnetwork.org/our-work/ ecological-footprint/

75 "Consumer Spending Trends and Current Statistics," Kimberly Amadeo, *The Balance*, June 27, 2019. https://www.thebalance.com/consumer-spending-trends-and-current-statistics-3305916 Also:

"Consumer Spending and the Economy," Hale Stewart, *The New York Times*, September 9, 2010. "The U.S. economy is predominantly driven by consumer spending, which accounts for approximately 70 % of all economic growth. But if consumers are to continue to drive the economy, they must be in a sound financial position; if they become overburdened with debt, they are not able to maintain their position as the primary driver of economic growth." https://fivethirtyeight.blogs.nytimes.com/2010/09/19/consumer-spending-and-the-economy/

76 "Climate change: Big lifestyle changes 'needed to cut emissions'", Roger Harrabin, *BBC*, August 2019. https://www.bbc.com/news/science-environment-49499521

77 Katherine Rooney, "Climate change will shrink these economies fastest," *World Economic Forum*, September 30, 2019. https://www.weforum.org/ agenda/2019/09/climate-change-shrink-these-economies-fastest/

78 Nicholas Stern, "Climate change will force us to redefine economic growth," *World Economic Forum*, July 11, 2018.

https://www.weforum.org/agenda/2018/07/here-are-the-economi
c-reasons-to-act-on-climate-change-immediately/

79   Paul Buchheit, "These 6 Men Have as Much Wealth as Half the World's
     Population," *Common Dreams*, February 20, 2017. https://www.ecowatch.
     com/richest-men-in-the-world-2274065153.html

80   "Oxfam says wealth of richest 1% equal to other 99%." *BBC*, January 16,
     2016. https://www.bbc.com/news/business-35339475

81   David Leonhardt, "The Rich Really Do Pay Lower Taxes Than You,"
     The New York Times, October 6, 2019. https://www.nytimes.
     com/interactive/2019/10/06/opinion/income-tax-rate-wealthy.
     html?action=click&module=Opinion&pgtype=Homepage

82   Jason Hickel, "Global inequality may be much worse than we think," *The
     Guardian*, April 8, 2016. "Global inequality is worse than at any time since
     the 19th century. . . It doesn't matter how you slice it; global inequality is
     getting worse. Much worse. Convergence theory turned out to be wildly
     incorrect. Inequality doesn't disappear automatically; it all depends on
     the balance of political power in the global economy. As long as a few
     rich countries have the power to set the rules to their own advantage,
     inequality will continue to worsen." https://www.theguardian.com/
     global-development-professionals-network/2016/apr/08/global-inequalit
     y-may-be-much-worse-than-we-think

83   "Extreme Carbon Inequality," Ibid.

84   "Climate Justice," *Wikipedia*, https://en.wikipedia.org/wiki/Climate_justice

85   Andrew Hoerner and Nia Robinson, "A Climate of Change: African
     Americans, Global Warming, and a Just Climate Policy for the US,"
     Environmental Justice & Climate Change Initiative, 2008. https://www.
     reimaginerpe.org/cj/hoerner-robinson

86   The "champagne glass" representation of inequities is commonly derived
     from the United Nations Human Development Report published in
     1992 and published in the Oxford University Press, 1992. Also see: Isabel
     Ortiz and Matthew Cummins, "Global Inequality: Beyond the Bottom
     Billion," UNICEF, Working Paper, April, 2011. See Figure 7, "Global Income
     Distributed by Percentiles of the Population in 2007," page 21. https://www.
     unicef.org/socialpolicy/files/Global_Inequality_REVISED_-_5_July.pdf
     Also see the "UNDESA World Social Report for 2020." https://www.un.org/
     development/desa/dspd/world-social-report/2020-2.html

87   I recognize how this terminology can be problematic because it assumes
     that the direction that currently "developed" nations have taken (toward
     overconsumption and hyper-individualization) is the agreed upon goal,
     and that "developing" nations are simply lagging in their achievement of
     that goal.

88   "Scientific Consensus: Earth's Climate is Warming," *NASA: Global Climate
     Change, Vital Signs of the Planet*, 2019. See the evidence here:

https://climate.nasa.gov/evidence/ See scientific consensus regarding climate warming here: https://climate.nasa.gov/scientific-consensus/ Also:

"Climate change: Disruption, risk and opportunity," *Science Direct* (originally published in *Global Transitions*, Volume 1, 2019. The study concludes: Climate change is disruptive because humans have adapted to a narrow range of environmental conditions. Change is particularly risky in the presence of low predictability, large scale, rapid onset and lack of reversibility. https://doi.org/10.1016/j.glt.2019.02.001

"Global Warming Science: The science is clear. Global warming is happening." *Union of Concerned Scientists*, 2019. https://www.ucsusa.org/our-work/global-warming/science-and-impacts/global-warming-science

89 Timothy M. Lenton, et. al., "Climate tipping point—too risky to bet against," *Nature*, November 27, 2019. https://www.nature.com/articles/d41586-019-03595-0 Also:

Arthur Neslen, "By 2030, We Will Pass the Point Where We Can Stop Runaway Climate Change," *HuffPost*, September 5, 2018, https://www.huffingtonpost.com/entry/runaway-climate-change-2030-report_us_5b8ecba3e4b0162f4727a09f

The 2030s may be a period of high instability in climate trends. Perhaps a climatological "whiplash." For example, a 2015 study predicted cooling rather than warming in this decade: "Solar activity predicted to fall 60% in 2030s, to mini-ice age levels: Sun driven by double dynamo," July 9, 2015, *Royal Astronomical Society*, reported in *Science Daily*. https://www.sciencedaily.com/releases/2015/07/150709092955.htm

Alexander Robinson, et al., "Multistability and critical thresholds of the Greenland ice sheet," *Nature Climate Change*, March 1, 2012. ". . . the Greenland ice sheet is more sensitive to long-term climate change than previously thought. We estimate that the warming threshold leading to an essentially ice-free state is in the range of $0.8-3.2°C$, with a best estimate of $1.6°C$." https://www.nature.com/articles/nclimate1449#citeas

Michael Marshall, "Major methane release is almost inevitable," *New Scientist*, February 21, 2013. "We are on the cusp of a tipping point in the climate. If the global climate warms another few tenths of a degree, a large expanse of the Siberian permafrost will start to melt uncontrollably." https://www.newscientist.com/article/dn23205-major-methane-release-is-almost-inevitable/#ixzz5zQ199XTi

"Jessica Corbett, "'Boiling with methane': Scientists reveal 'truly terrifying' sign of climate change under the Arctic Ocean," *Common Dreams*, October 9, 2019. https://www.alternet.org/2019/10/boiling-with-methane-scientists-reveal-truly-terrifying-sign-of-climate-change-under-the-arctic-ocean/

90 "Temperature rise is 'locked-in' for the coming decades in the Arctic," *UNEP*, March 12, 2019. "Even if existing Paris Agreement commitments are met, winter temperatures over the Arctic Ocean will increase $3-5°C$ by mid-century compared to 1986-2005 levels. Thawing permafrost could wake

'sleeping giant' of more greenhouse gases, potentially derailing global climate goals." https://www.unenvironment.org/news-and-stories/press-release/temperature-rise-locked-coming-decades-arctic Also:

Steffen, et. al., "Trajectories of the Earth System in the Anthropocene," *PNAS*, July 6, 2018. This study explores: Hothouse Earth and how runaway global warming threatens 'habitability of the planet for humans. https://www.pnas.org/content/115/33/8252?mod=article_inline

91 "An unexpected surge in global atmospheric methane is threatening to erase the anticipated gains of the Paris Climate Agreement. Previously stable global methane levels have unexpectedly surged in recent years. See: Benjamin Hmiel, et.al., "Preindustrial $^{14}CH_4$ indicates greater anthropogenic fossil CH_4 emissions," *Nature*, February 19, 2020. https://www.nature.com/articles/s41586-020-1991-8 This study shows that scientists and governments have been greatly underestimating emissions of the powerful greenhouse gas methane from oil and gas operations. Also:

Nisbet et al. "Very Strong Atmospheric Methane Growth in the 4 Years 2014–2017: Implications for the Paris Agreement," *Global Biogeochemical Cycles*. March 2019. https://doi.org/10.1029/2018GB006009 See the summary article in *Climate Nexus* here: https://climatenexus.org/climate-change-news/methane-surge/

92 Hubau Wannes, et al., "Asynchronous carbon sink saturation in African and Amazonian tropical forests," *Nature*, March 5, 2020. https://www.nature.com/articles/s41586-020-2035-0 Also:

Fiona Harvey, "Tropical forests losing their ability to absorb carbon, study finds," *The Guardian*, March 4, 2020. https://www.theguardian.com/environment/2020/mar/04/tropical-forests-losing-their-ability-to-absorb-carbon-study-finds

93 Stewart Patrick, "The Coming Global Water Crisis," *The Atlantic*, May 9, 2012. https://www.theatlantic.com/international/archive/2012/05/the-coming-global-water-crisis/256896/ Also:

William Wheeler, "Global water crisis: too little, too much, or lack of a plan?," *Christian Science Monitor*, December 2, 2012. https://www.csmonitor.com/World/Global-Issues/2012/1202/Global-water-crisis-too-little-too-much-or-lack-of-a-plan

94 Gilbert Houngbo, "The United Nations world water development report 2018: nature-based solutions for water," UNESCO, 2018. https://unesdoc.unesco.org/ark:/48223/pf0000261424

95 Stephen Leahy, "From Not Enough to Too Much, the World's Water Crisis Explained," *National Geographic*, March 22, 2018. https://www.nationalgeographic.com/news/2018/03/world-water-day-water-crisis-explained/

96 Paul Salopek, "Historic water crisis threatens 600 million people in India," *National Geographic*, October 19, 2018. https://www.nationalgeographic.

com/culture/water-crisis-india-out-of-eden/?cmpid=org=ngp::mc=crm-
email::src=ngp::cmp=editorial::add=Science_20200129&rid=51139F7FFE
E4083137CDD6D1FF5C57FF

97   Dan Charles, "5 Major Crops In The Crosshairs Of Climate Change," *NPR*,
     October 25, 2018. https://www.npr.org/sections/thesalt/2018/10/25/65858
     8158/5-major-crops-in-the-crosshairs-of-climate-change Also:

     Sean Illing, "The climate crisis and the end of the golden era of food choice,"
     Vox, June 24, 2019. https://www.vox.com/the-highlight/2019/6/17/186341
     98/food-diet-climate-change-amanda-little

     Rachel Nuwer, "Here's how climate change will affect what you eat," *BBC*,
     December 28, 2015. https://www.bbc.com/future/article/20151228-heres-
     how-climate-change-will-affect-what-you-eat

     Nicholas Thompson, "The Most Delicious Foods Will Fall Victim to Climate
     Change," *Wired*, June 13, 2019. https://www.wired.com/story/the-most-
     delicious-foods-will-fall-victim-to-climate-change/

     Ian Burke, "29 of Your Favorite Foods That Are Threatened by
     Climate Change," *Saveur*, June 7, 2017. https://www.saveur.com/
     climate-change-ingredients/

     Daisy Simmons, "A brief guide to the impacts of climate change on food
     production," *Yale Climate Connections*, September 18, 2019. https://www.
     yaleclimateconnections.org/2019/09/a-brief-guide-to-the-impacts-of-
     climate-change-on-food-production/

     Ilima Loomis "Get ready to eat differently in a warmer world," *Science News
     for Students*, May 23, 2019. https://www.sciencenewsforstudents.org/
     article/climate-change-global-warming-food-eating

     Peter Schwartzstein, "Indigenous farming practices failing as climate change
     disrupts seasons," *National Geographic*, October 14, 2019. https://www.
     nationalgeographic.com/science/2019/10/climate-change-killing-thousand
     s-of-years-indigenous-wisdom/

     Kay Vandette, "Climate change could make leafy greens, veggies
     less available," *Earth*, June 11, 2018. https://www.earth.com/news/
     climate-change-could-make-leafy-greens-veggies-less-available/

98   Current World Population: https://www.worldometers.info/
     world-population/

99   "Nature's Dangerous Decline 'Unprecedented'; Species Extinction Rates
     'Accelerating'" *Intergovernmental Science-Policy Platform on Biodiversity
     and Ecosystem Services (IPBES)*, May 22, 2019. https://www.ipbes.net/
     news/Media-Release-Global-Assessment

100 "Ocean Deoxygenation," *International Union for Conservation of Nature*,
     December 8, 2019. Marine life, fisheries increasingly threatened as the
     ocean loses oxygen. Even the smallest fall in oxygen levels, when near
     already existing thresholds, can create significant issues with far-reaching

and complex biological and biogeochemical implications. https://www.iucn.org/theme/marine-and-polar/our-work/climate-change-and-oceans/ocean-deoxygenation

101 Adapted from John Fullerton, "Regenerative Capitalism How Universal Principles And Patterns Will Shape Our New Economy," *Capital Institute*, April 2015. https://capitalinstitute.org/wp-content/uploads/2015/04/2015-Regenerative-Capitalism-4-20-15-final.pdf?mc_cid=236080d2f0&mc_eid=2f41fb9d8d

102 "Richest 1% on target to own two-thirds of all wealth by 2030," *The Guardian*, April 7, 2018. https://www.theguardian.com/business/2018/apr/07/global-inequality-tipping-point-2030

103 Duane Elgin, "Limits to Complexity: Are Bureaucracies Becoming Unmanageable," *The Futurist,* December 1977. https://duaneelgin.com/wp-content/uploads/2014/11/Limits-to-Large-Complex-Systems.pdf

104 "Transitions and Tipping Points in Complex Environmental Systems,"A Report by the *National Science Foundation Advisory Committee for Environmental Research and Education*, 2009. https://www.nsf.gov/ere/ereweb/ac-ere/nsf6895_ere_report_090809.pdf This is not a time specific warning but rather a general one from 2009: "The world is at a crossroads. The global footprint of humans is such that we are stressing natural and social systems beyond their capacities. We must address these complex environmental challenges and mitigate global scale environmental change or accept likely all-pervasive disruptions. . . The rate of environmental change is outpacing the ability of institutions and governments to respond effectively."

105 T. Schuur, "Arctic Report Card: Permafrost and the Global Carbon Cycle," *NOAA*, 2019. https://arctic.noaa.gov/Report-Card/Report-Card-2019/ArtMID/7916/ArticleID/844/Permafrost-and-the-Global-Carbon-Cycle

106 "Fighting Wildfires Around the World," *Frontline, Wildfire Defense Systems*, 2019. https://www.frontlinewildfire.com/fighting-wildfires-around-world/

107 Carrying capacity estimates, Op. Cit.

108 Iliana Paul, "Climate Change and Social Justice," *WEDO,* 2014. https://www.wedo.org/wp-content/uploads/wedo-climate-change-social-justice.pdf?utm_source=newsletter&utm_medium=email&utm_content=http%3A//d31hzlhk6di2h5.cloudfront.net/20161107/ce/11/85/a8/5d76d1fbe015e871ef155f93_386x486.png&utm_campaign=Emma%20Newsletter

109 Dmitry Orlov, *Reinventing Collapse: The Soviet Example and American Prospects*, New Society Publishers, 2008.

Also see: Tainter, *The Collapse of Complex Societies*, Op. Cit.

110 Carrying capacity estimates, Op. Cit.

111 Op. Cit., "Nature's Dangerous Decline 'Unprecedented'; Species Extinction Rates 'Accelerating'" *Intergovernmental Science-Policy Platform on*

Biodiversity and Ecosystem Services (IPBES), May 22, 2019.
https://www.ipbes.net/news/Media-Release-Global-Assessment

112 Plants are also likely to experience the stress and trauma of the great
dying. See Nicoletta Lanese, "Plants 'Scream' in the Face of Stress,"
Live Science, December 6, 2019. https://www.livescience.com/
plants-squeal-when-stressed.html

113 My assessment that several billion people may perish in the latter portion
of the time frame of this scenario (where the world is not powered by
fossil fuels) has been described as hyper-optimistic. Jason Brent (http://
www.jgbrent.com/about-the-author.html) considers it likely that many
more will die. See his reply to my article, "Existential threats, Earth Voice
and the Great Transition," *Millennium Alliance for Humanity and the
Biosphere,* MAHB, January 21, 2020. https://mahb.stanford.edu/blog/
mahb-dialogue-author-humanist-duane-elgin/ Brent writes: "The collapse of
civilization will occur because humanity is in overshoot using the resources
of 1.7 Earth's and going deeper into overshoot every second due to a growing
population (expected to grow by 3.2 billion reaching 10.9 billion by the year
2100—a 41.5% growth in 80 years) and a growing per capita worldwide
usage of resources. Simple math shows that to get out of overshoot the
human population would have to be reduced to 4.47 billion. If population
were to reach 10.9 billion, that would require a reduction in population of
6.43 billion (10.9-4.47= 6.43) (without considering any reduction due to the
per capita increase in the usage of resources) to get out of overshoot. Simple
statement—there is zero chance that voluntary population control will
achieve that reduction (of 6.3 billion) prior to the collapse of civilization and
the deaths of billions."

114 The Great Burning began in 2019. See: Laura Paddison, "2019 Was The Year
The World Burned," *HuffPost*, December 27, 2019. https://www.huffpost.
com/entry/wildfires-california-amazon-indonesia-climate-change_n_
5dcd3f4ee4b0d43931d01baf Also:

At least a billion animals are estimated to have died by 2020 in the bushfires
in Australia. Lisa Cox, "A billion animals: some of the species most at risk
from Australia's bushfire crisis," *The Guardian*, Jan 13, 2020. The ecologist
Chris Dickman has estimated more than a billion animals have died around
the country—a figure that excludes fish, frogs, bats and insects. . . "This is
just the tip of the iceberg," James Trezise, a policy analyst at the Australian
Conservation Foundation, says. "The number of species and ecosystems
that have been severely impacted across their ranges is almost certain to be
much higher, especially when factoring in less well-known species of reptiles,
amphibians and invertebrates." https://www.theguardian.com/australia-
news/2020/jan/14/a-billion-animals-the-australian-species-most-at-risk-
from-the-bushfire-crisis

The great burning to come is powerfully summarized in the following video
that shows a woman rescuing a badly burnt and wailing koala bear from an
Australian bushfire. The marsupial was spotted crossing a road among the
flames. A local woman rushed to the koala's aid, wrapping the animal in her

shirt and a blanket and pouring water over it. She took the injured animal to a nearby koala hospital. It is truly heart breaking to watch innocents suffer for reasons not their own—and to realize this is our future unless we respond swiftly. https://www.youtube.com/watch?v=3x8JXQ6RTIU

115 "Wildfires in the Amazon are predicted to worsen, doubling the affected area of an important part of the forest by 2050. The result could be to shift the Amazon from a carbon sink into a net source of carbon dioxide emissions," See story: "Burning of Amazon may get a lot worse," *New Scientist*, January 18, 2020. Also:

Herton Escobar, "Brazil's deforestation is exploding—and 2020 will be worse," *Science Magazine*, November 22, 2019. https://www.sciencemag.org/news/2019/11/brazil-s-deforestation-exploding-and-2020-will-be-worse?utm_campaign=news_daily_2019-11-22&et_rid=510705016&et_cid=3086753

116 Stephen Pune, "California wildfires signal the arrival of a planetary fire age," *The Conversation*, November 1, 2019. https://theconversation.com/california-wildfires-signal-the-arrival-of-a-planetary-fire-age-125972

117 John Pickrell, "Massive Australian blazes will 'reframe our understanding of bushfire,'" *Science Magazine*, November 20, 2019. https://www.sciencemag.org/news/2019/11/massive-australian-blazes-will-reframe-our-understanding-bushfire?utm_campaign=news_daily_2019-11-20&et_rid=510705016&et_cid=3083308 Also:

Damien Cave, "Australia Burns Again, and Now Its Biggest City Is Choking," *The New York Times,* December 6, 2019. https://www.nytimes.com/2019/12/06/world/australia/sydney-fires.html A

118 Stephen Pyne, "The Planet is Burning," *Aeon*, November 2019. Also:

Stephen Pyne, *Fire: A Brief History* (2019). https://aeon.co/essays/the-planet-is-burning-around-us-is-it-time-to-declare-the-pyrocene

David Wallace-Wells, "In California, Climate Change Has Turned Rainy Season Into Fire Season," *New York Magazine*, November 12, 2018. https://nymag.com/intelligencer/2018/11/the-california-fires-and-the-threat-of-climate-change.html Edward Helmore, "'Unprecedented': more than 100 Arctic wildfires burn in worst ever season," *The Guardian*, July 26, 2019. The article describes, "Huge blazes in Greenland, Siberia and Alaska are producing plumes of smoke that can be seen from space." https://www.theguardian.com/world/2019/jul/26/unprecedented-more-than-100-wildfires-burning-in-the-arctic-in-worst-ever-season

119 Hans Seyle, was a highly regarded endocrinologist known for his studies of the effects of stress on the human body. https://www.azquotes.com/author/13308-Hans_Selye

120 Francis Weller, *The Wild Edge of Sorrow*, North Atlantic Books, 2015.

121 Weller, Ibid.

122 Naomi Shihab Nye, *Words Under the Words: Selected Poems*, 1995. https://poets.org/poem/kindness

123 "Global Cities at Risk from Sea Level Rise: Google Earth Video," *Climate Central*, 2019. https://sealevel.climatecentral.org/maps/google-earth-video-global-cities-at-risk-from-sea-level-rise Also:

Scott Kulp, et al., "New elevation data triple estimates of global vulnerability to sea-level rise and coastal flooding," *Nature Communications*, October 29, 2019. Some of the earlier projections of population displacement from sea-level rise are probably way too low. Around the world, instead of some 50 million people being forced to move to higher ground over the next 30 years, the oceans will likely rise higher than predicted, with a coastal diaspora at least three times larger; by 2100, the number of climate refugees could surpass 300 million. https://www.nature.com/articles/s41467-019-12808-z Other estimates place the number of climate refugees as high as 2 billion people by 2100.

Charles Geisler & Ben Currens, "Impediments to inland resettlement under conditions of accelerated sea level rise," *Land Use Policy*, March 29, 2017. The authors extrapolate from 2060 to conclude that in the year 2100, 2 billion people—about one-fifth of a world population of 11 billion—could become climate change refugees due to rising ocean levels. https://doi.org/10.1016/j.landusepol.2017.03.029

Blaine Friedlander, "Rising seas could result in 2 billion refugees by 2100," *Cornell Chronicle*, June 19, 2017. http://news.cornell.edu/stories/2017/06/rising-seas-could-result-2-billion-refugees-2100

124 Jennifer Welwood, "The Dakini Speaks," http://jenniferwelwood.com/poetry/the-dakini-speaks/

125 Todd May, "Would Human Extinction Be a Tragedy?" *The New York Times*, December 17, 2018. https://www.nytimes.com/2018/12/17/opinion/human-extinction-climate-change.html

126 Wallace Stevens, *Goodreads*, https://www.goodreads.com/quotes/565035-after-the-final-no-there-comes-a-yes-and

127 To illustrate the difficulty of meeting the net zero CO_2 emission goals by 2050, see the *World Energy Outlook 2019* which concludes the world's CO_2 emissions are set to continue rising for decades unless there is greater ambition on climate change, despite the "profound shifts" already underway in the global energy system. That is one of the key messages from the International Energy Agency's (IEA) World Energy Outlook 2019. https://webstore.iea.org/world-energy-outlook-2019

128 Joanna Macy and Chris Johnstone, *Active Hope: How to Face the Mess We're in Without Going Crazy*, New World Library, 2012.

129 Of great concern is when cumulative global CO_2 emissions exceed the 1 trillion tons of carbon threshold, which according to the IPCC will raise the Earth's surface temperature to 2°C above the pre-industrial minimum and

trigger "dangerous interference" with the Earth's climate system. When will the 1 trillion-ton threshold be exceeded? Estimates are sometime between 2050 and 2055 regardless of which population growth scenario is used. "Global CO_2 emissions forecast to 2100," Roger Andrews, *Euanmearns*, March 7, 2018. http://euanmearns.com/global-energy-forecast-to-2100/

130 Impacts of a 4 degree Celsius Global Warming, Green Facts, https://www. greenfacts.org/en/impacts-global-warming/l-2/index.htm Also:

A broad consensus exists that 4° Celsius will happen by the end of the century or before if no major actions are taken. "Climate change may be escalating so fast it could be 'game over,' scientists warn." A climate range between 4.8 C and 7.4 C by 2100 emerged from calculations published in the journal, *Science Advances*. https://advances.sciencemag.org/content/2/11/e1501923

Ian Johnston, "Climate change may be escalating so fast it could be 'game over,' scientists warn." *Independent*, November 9, 2016. https://www. independent.co.uk/news/science/climate-change-game-over-global-warm ing-climate-sensitivity-seven-degrees-a7407881.html

David Wallace-Wells, "UN says climate genocide is coming," *New York Magazine,* October 10, 2019. He states that the planet is on a trajectory that "brings us north of four degrees by the end of the century." http://nymag. com/intelligencer/2018/10/un-says-climate-genocide-coming-but-its-worse-than-that.html

Roger Andrews, "Global CO_2 emissions forecast to 2100," Blog by Euan Mearns, March 7, 2018. http://euanmearns.com/global-CO2-emissions-forecast-to-2100/

"4 Degrees Hotter, A Climate Action Centre Primer," *Climate Code Red*, February 2011. Melbourne, Australia. https://www.climatecodered.org/20 11/02/4-degrees-hotter-adaptation-trap.html The study quotes Professor Kevin Anderson, director of the Tyndall Centre for Climate Change, who "believes only around 10% of the planet's population—around half a billion people—will survive if global temperatures rise by 4 C. He said the consequences were "terrifying." "For humanity it's a matter of life or death," he said. "We will not make all human beings extinct as a few people with the right sort of resources may put themselves in the right parts of the world and survive. But I think it's extremely unlikely that we wouldn't have mass death at 4 C." In 2009 Professor Hans Joachim Schellbhuber, director of the Potsdam Institute, and one of Europe's most eminent climate scientists, told his audience that at 4 C, population ". . . carrying capacity estimates (are) below 1 billion people." p. 9.

Another estimate of the carrying capacity of the Earth is found in the magazine, *New Scientist,* November1, 2014, p. 9. Corey Bradshaw and Barry Brook, Op. Cit., suggest that a sustainable human population, given current Western consumption patterns and technologies, would be between 1 and 2 billion people.

131 The researchers used the MIT Integrated Global System Model Water Resource System (IGSM-WRS) to evaluate water resources and needs worldwide. See: "Water Stress to Affect 52% of World's Population by 2050," *Water Footprint Network*, https://waterfootprint.org/en/about-us/news/news/water-stress-affect-52-worlds-population-2050/

132 Op. Cit. The United Nations world water development report 2018: nature-based solutions for water. Also:

Claire Bernish, "Water Scarcity Will Make Life Miserable for Nearly 6 Billion People by 2050," *The Mind Unleashed*, March 23, 2018. https://themindunleashed.com/2018/03/water-scarcity-6-billion-2050.html More than 5 billion people could suffer water shortages by 2050 due to climate change, increased demand and polluted supplies, according to a UN report on the state of the world's water. Without drastic changes focused on natural solutions nearly six billion people will be in the grips of a punishing water shortage by 2050.

133 Joseph Hinks, "The World Is Headed for a Food Security Crisis," *TIME* magazine, March 28, 2018. https://time.com/5216532/global-food-security-richard-deverell/

134 Rebecca Chaplin-Kramer, "Global modeling of nature's contributions to people," *Science*, Vol. 366, Issue 6462, October 11, 2019. https://science.sciencemag.org/content/366/6462/255 Also:

Miyo McGinn, "New study pinpoints the places most at risk on a warming planet," *Grist*, October 17, 2019. https://grist.org/article/new-study-pinpoints-the-places-most-at-risk-on-a-warming-planet/

135 Francois Gemenne, "A review of estimates and predictions of people displaced by environmental changes," Global Environmental Change, in *Science Direct*, December 2011. https://www.sciencedirect.com/science/article/abs/pii/S0959378011001403?via%3Dihub

136 Worldometers: https://www.worldometers.info/world-population/

137 See, for example, Op. Cit., Ishan Daftardar, "Why Bee Extinction Would Mean the End of Humanity," *Science ABC,* July 23, 2015. https://www.scienceabc.com/nature/bee-extinction-means-end-humanity.html

138 J. Hempel, "Social Media Made the Arab Spring, But Couldn't Save It," *Wired*, January 26, 2016. https://www.wired.com/2016/01/social-media-made-the-arab-spring-but-couldnt-save-it/

139 "Russia 'meddled in all big social media' around US election," BBC, December 17, 2018. https://www.bbc.com/news/technology-46590890

140 Charles Geisler & Ben Currens, "Impediments to inland resettlement under conditions of accelerated sea level rise," *Land Use Policy*, March 29, 2017. The authors extrapolate from 2060 to conclude that in the year 2100, 2 billion people—about one-fifth of a world population of 11 billion—could become climate change refugees due to rising ocean levels. https://doi.org/10.1016/j.landusepol.2017.03.029

141 Chris Mooney, "Unprecedented data confirms that Antarctica's most dangerous glacier is melting from below," *Washington Post*, January 30, 2020. "Warm ocean water has been discovered underneath a massive glacier in West Antarctica, a troubling finding that could speed its melt in a region with the potential to eventually unleash more than 10 feet of sea-level rise." David Holland, a New York University glaciologist said the warm water melting the glacier from below is, "really, really bad . . . That's not a sustainable situation for that glacier." "Researchers believe that as recently as some 100,000 years ago, West Antarctica was not a sheet of ice at all—but rather, an open ocean that later converted to glacier. The fear is that the melting now taking place could lead to a return to open ocean." https://www.washingtonpost.com/climate-environment/2020/01/30/unprecedented-data-confirm-that-antarcticas-most-dangerous-glacier-is-melting-below/ Also:

"Thwaites Glacier," *The International Thwaites Glacier Collaboration.* https://thwaitesglacier.org/about/itgc

"Thwaites Glacier Transformed," *NASA Earth Observatory*, December 28, 2019. ". . . urgency stems from observations and analyses showing that the amount of ice flowing from Thwaites—and contributing to sea level rise—has doubled in the span of three decades. Scientists think the glacier could undergo even more dramatic changes in the near future." https://earthobservatory.nasa.gov/images/146247/thwaites-glacier-transformed

142 Fred Pearce, "Geoengineer the Planet? More Scientists Now Say It Must Be an Option," Yale School of Forestry & Environmental Studies, May 29, 2019. https://e360.yale.edu/features/geoengineer-the-planet-more-scientists-now-say-it-must-be-an-option Climate engineering or climate intervention, commonly referred to as geoengineering, is the deliberate and large-scale intervention in the Earth's climate system, usually with the aim of mitigating the adverse effects of global warming. When considering which is the *worse* option, either geoengineering of climate or unrestrained global warming, it seems that the geoengineering approach may be the better option. There is wide range of proposed techniques. Generally, these can be grouped into two categories: 1) **solar** geoengineering to reflect a small proportion of the Sun's energy back into space, counteracting the temperature rise caused by increased levels of greenhouse gases in the atmosphere which absorb energy and raise temperatures. and 2) **carbon** geoengineering that removes carbon dioxide or other greenhouse gases from the atmosphere, directly countering the increased greenhouse effect and ocean acidification. These techniques have to be implemented on a global scale to have a significant impact on greenhouse gas levels. A climate expert, Fred Pearce, writes that perhaps the most promising answer lies in going back to nature's systems and restoring natural forests. Also:

"What Is Geoengineering?" *Oxford Geoengineering Program*, 2018. http://www.geoengineering.ox.ac.uk/www.geoengineering.ox.ac.uk/what-is-geoengineering/what-is-geoengineering/

Christopher H. Trisos et al., "Potentially dangerous consequences for biodiversity of solar geoengineering implementation and termination,"

Nature Ecology and Evolution, January 22, 2018. https://www.nature.com/articles/s41559-017-0431-0 "Geoengineering could cause more harm than climate change," *Cosmos Magazine*, January 23, 2018. https://cosmosmagazine.com/technology/neering-could-cause-more-harm-than-climate-change

143 Martin Luther King, Jr. quoted in Stephen B. Oates, *Let the Trumpets Sound: The Life of Martin Luther King, Jr.*, New American Library, 1982.

144 T.S. Eliot, "Four Quartets, Little Gidding," 1943. https://www.brainyquote.com/quotes/t_s_eliot_109032

145 Simone de Beauvoir, See "Brainy Quotes": https://www.brainyquote.com/quotes/simone_de_beauvoir_392724

146 Joseph Campbell, et al., "Changing Images of Man," *Center for the Study of Social Policy, Stanford Research Institute*, Menlo Park, California. The study was prepared for the Kettering Foundation, Dayton, Ohio, Contact: URH (489)-2150, May 1974 and subsequently republished with the same title in 1982 by Pergamon Press.

147 Joseph Campbell & Bill Moyers, *The Power of Myth*, Archer, 1988. https://www.goodreads.com/quotes/10442-people-say-that-what-we-re-all-seeking-is-a-meaning

148 I am grateful for the wisdom of Erich Fromm and his description: "On the Wisdom and the Meaning of Life," *Excellence Reporter*, May 28, 2019. I have revised his insights into my own words to reframe his masculine emphasis. https://excellencereporter.com/2019/05/28/erich-fromm-on-the-wisdom-and-the-meaning-of-life/

149 Florida Scott-Maxwell, *The Measure of My Days*, Penguin Books, 1979. https://www.goodreads.com/author/quotes/550910.Florida_Scott_Maxwell

150 To learn into our time of great transition and beyond, my partner Coleen and I have convened a learning community of roughly several dozen people in the past year. Our collective explorations have been very valuable in grounding the work described in this book.

151 Malcolm Margolin, *The Ohlone Way: Indian Life in the San Francisco-Monterey Bay Area*, Berkeley: Heyday Books, 1978.

152 Richard Nelson, *Make Prayers to the Raven*, Chicago: University of Chicago Press, 1983.

153 Luther Standing Bear, quoted in J.E. Brown, "Modes of contemplation through actions: North American Indians." In *Main Currents in Modern Thought*, New York, November-December 1973.

154 Anne Baring, Op. Cit., p. 83.

155 Ann Baring, Op. Cit., p. 421.

156 Mathew Fox, *Meditations with Meister Eckhart*, Santa Fe, NM: Bear & Co., 1983.

157 See, for example, Coleman Barks, *The Essential Rumi*, San Francisco: Harper San Francisco, 1995.

158 D.T. Suzuki, *Zen and Japanese Culture*, Princeton, NJ: Princeton University Press, 1970.

159 S. N. Maharaj, *I Am That*. Part I (trans., Maurice Frydman), Bombay, India: Chetana, 1973.

160 Lao Tsu, *Tao Te Ching* (trans. Gia-Fu Feng and Jane English), New York: Vintage Books, 1972.

161 See, for example: https://www.goodreads.com/quotes/tag/mysticism Also: http://www.gardendigest.com/myst1.htm

162 Henry Thoreau, https://www.goodreads.com/quotes/32955-heaven-is-under-our-feet-as-well-as-over-our

163 Predrag Cicovacki, *Albert Schweitzer's Ethical Vision A Sourcebook*, Oxford University Press, February 2, 2009.

164 Gary Comstock and Susan E. Henking, *Que(e)rying Religion: A Critical Anthology*, Continuum, February 1, 1997. https://www.amazon.com/s?k=9780826409249&i=stripbooks&linkCode=qs

165 John Muir, https://www.adventure-journal.com/2013/10/the-aj-list-20-inspiring-quotes-from-john-muir/

166 John Muir, https://www.passiton.com/inspirational-quotes/7869-and-into-the-forest-i-go-to-lose-my-mind-and

167 Gary Snyder, https://www.goodreads.com/quotes/119749-nature-is-not-a-place-to-visit-it-is-home

168 Haruki Murakami, https://www.goodreads.com/quotes/448426-not-just-beautiful-though-the-stars-are-like-the

169 Walt Whitman, https://poets.org/poem/song-myself-31

170 Joseph Campbell, https://www.brainyquote.com/quotes/joseph_campbell_387298

171 Buddha, https://www.spiritualityandpractice.com/quotes/quotations/view/198/spiritual-quotation

172 John Muir, https://www.goodreads.com/quotes/40749-one-touch-of-nature-makes-the-whole-world-kin

173 Frank Lloyd Wright, https://www.brainyquote.com/quotes/frank_lloyd_wright_107515

174 Mira Michelle, "The Meaning of Life . . . Communing Deeply with Nature," *Excellence Reporter*, April 15, 2019. https://excellencereporter.com/2019/04/15/miramichelle-the-meaning-of-life-communing-deeply-with-nature/

175 George Washington Carver, http://www.quotes-inspirational.com/quote/you-love-enough-anything-talk-78/

176 Laura Marie Edinger-Schons, "People with a sense of oneness, experience greater life satisfaction. Effect is found regardless of religion," *American Psychological Association*, April 11, 2019. https://www.sciencedaily.com/releases/2019/04/190411101803.htm The surveys revealed that people who have a feeling of oneness—a feeling that everything in the world is connected and interdependent—have greater life satisfaction than those who do not, regardless of whether they belong to a religion or not. Also:

John de Castro, "A model of enlightened/mystical/awakened experience," *Psychology of Religion and Spirituality*, (American Psychological Association), February 2017. https://psycnet.apa.org/doiLanding?doi=10.1037%2Frel0000037 The study found that awakening or transcendent experiences are powerful and profoundly affect the individual. There appears to be an essential core experience of oneness. It is experienced as a completely subjective phenomenon where awareness contains all of reality and the sense of a separate self is perceived as a delusion. The author hypothesizes that awakening experiences result from the progressive removal of the cognitive, perceptual, and sensory layers of information processing and returns awareness to a primal state that was present before the development of neural information processing.

Daniel Stancato, et al., "Awe, ideological conviction, and perceptions of ideological opponents," *APA PsycNet*, December 2, 2019. "These findings indicate that awe may . . . promote reduced dogmatism and increased perceptions of social cohesion." https://psycnet.apa.org/record/2019-46364-001

177 Sallie McFague, *The Body of God: An Ecological Theology,* Fortress Press, 1993.

Also: Richard Rohr, "The Universe is the Body of God," https://cac.org/the-universe-is-the-body-of-god-2019-03-07/

178 Cynthia Bourgeault, "Fullness of Life," September 11, 2017. http://cynthiabourgeault.org/2017/09/11/fullness-of-life-2/

179 E. C. Roehlkepartain, et al., "With their own voices: A global exploration of how today's young people experience and think about spiritual development", *Search Institute*, 2008. https://www.search-institute.org/wp-content/uploads/2018/02/with_their_own_voices_report.pdf

180 "Many Americans Mix Multiple Faiths," *Pew Research Center, Religion & Public Life,* December 9, 2009. Mystical experiences shown in third figure, which references 1962 survey reported by Gallup and presented in *Newsweek*, April, 2006 See: https://www.pewforum.org/2009/12/09/many-americans-mix-multiple-faiths/

Also: Andrew Greely and William McCready, "Are We a Nation of Mystics," in *The New York Times Magazine*, January 26, 1976.

181 "U.S. public becoming less religious," *Pew Research Center*, November 3, 2015. Survey results on regular experiences of "peace and sense of wonder." https://www.pewforum.org/2015/11/03/u-s-public-becoming-less-religious/

182 T. Clarke, et al., "Use of Yoga, Meditation, and Chiropractors Among U.S. Adults Aged 18 and Over," *National Center for Health Statistics*, November 2018. https://www.ncbi.nlm.nih.gov/pubmed/30475686

183 In the spirit of full disclosure, my personal understanding of an ecology of consciousness permeating the universe was developed and documented in a wide-ranging series of scientific experiments over a period nearly three years, from 1972 to 1975, at the Stanford Research Institute (now SRI International), in Menlo Park, California. Although my primary work at the time was as a senior social scientist in the futures group at SRI, for nearly 3 years, I was a consultant to NASA to explore a wide range of experiments regarding intuitive capacities in the engineering laboratory—often 3 days a week for 2 or 3 hour stretches, depending on the experiments at the time, and all with various forms of feedback. Experiments included "remote viewing" of diverse locations and technologies; clairvoyance with a random number generator; influencing movement of pendulum-clock measured with a laser beam; interacting with a magnetometer immersed in a container filled with liquid helium; pressing on a balance pan scale from outside a locked room; influencing plant growth with comparisons to a controlled group. I dropped out of these fascinating experiments in 1975 when they were taken over by the CIA and declared secret (this research apparently continued for another 20 years according to Freedom of Information Act; see: Hal Puthoff, "CIA-Initiated Remote Viewing Program at Stanford Research Institute," *Journal of Scientific Exploration, Vol.* 10, No. 1, 1996). Based on my experience in these scientific experiments, I concluded that *everyone* has an intuitive faculty and literal connection with the universe. An empathic connection with the cosmos is not restricted to a gifted few, it is an ordinary part of the functioning of the universe and is accessible to all. These experiments made it clear that we have barely begun to develop a literacy of consciousness utilizing sophisticated technologies to provide feedback (similar to learning with bio-feedback but instead with bio-cosmic feedback). These experiments demonstrated that our being does not stop at the edge of our skin but extends into and is inseparable from the unified universe. A description of selected SRI experiments can be found at: Russell Targ, Phyllis Cole, and Harold Puthoff, "Development of Techniques to Enhance Man/Machine Communication," *Stanford Research Institute*, Menlo Park, California, prepared for NASA, contract 953653 Under NAS7-100, June 1974. Also:

Harold Puthoff and Russell Targ, "A Perceptual Channel for Information Transfer Over Kilometer Distances," published in the *Proceedings of the I.E.E.E.* (Institute of Electrical and Electronics Engineers), vol. 64, no. 3, March 1976.

R. Targ and H. Puthoff, *Mind-Reach: Scientists Look at Psychic Ability*, Delacorte Press/Eleaonor Friede, 1977.

184 Happold, F.C. (1975). Quoted by A. Greeley and W. McCready, W., "Are we a nation of mystics?" *The New York Times Magazine.* January 26, 1975.

185 Duane Elgin, *The Living Universe*, Op., Cit. Another way to consider the issue of aliveness is to explore the operating characteristics of biological systems and see whether the universe exhibits similar capacities. Generally, a system must include at least four key capacities to be considered living: 1) **Metabolism**—the ability to break matter down as well as to synthesize it. From its formation, the universe has been synthesizing simple matter (helium and hydrogen) and converting it through supernova into carbon, nitrogen, oxygen and sulfur—essential constituents from which we are made. 2) **Self-regulation**—the ability to maintain stability in its operation. The universe has endured and evolved over billions of years as a unified system that produces self-organizing systems at every scale, from atomic to galactic, that can persist for billions of years. 3) **Reproduction**—the ability to create copies of itself. A number of cosmologists theorize that on the other side of black holes are white holes giving birth to new cosmic systems. 4) **Adaptation**—the ability to evolve and fit into changing environments. The universe has evolved over billions of years to produce systems of increasing complexity and coherence woven together into a self-consistent whole. Because these four criteria are found, not only in plants and animals but also in the functioning of the universe, it seems valid to describe the universe as a unique kind of living system.

186 Clara Moskowitz, "What's 96 Percent of the Universe Made Of? Astronomers Don't Know," *Space.com*, May 12, 2011. https://www.space.com/11642-dar k-matter-dark-energy-4-percent-universe-panek.html

187 Graphic showing composition of the universe in *The Imagineers Home*. https://www.theimagineershome.com/blog/dark-matter-and-energy- linked-to-quantum-mechanics/

188 Brian Swimme, *The Hidden Heart of the Cosmos*, Orbis Books, May 1996. https://www.amazon.com/Hidden-Heart-Cosmos-Humanity-Ecology/ dp/1626983437

189 Freeman Dyson, https://en.wikiquote.org/wiki/Freeman_Dyson

190 Max Planck, Interview in *The Observer*, January 25, 1931. https://en.wikiquote.org/wiki/Max_Planck

191 John Gribbin, *In the Beginning: The Birth of the Living Universe*, New York: Little Brown, 1993.

Also see: David Shiga, "Could black holes be portals to other universes?" *New Scientist*, April 27, 2007.

192 Albert Einstein, *Ideas and Opinions*, New York: Bonanza Books,1954.

193 Albert Einstein's famous quote was written in 1950 in a letter to Robert S. Marcus, a man who was distraught over the death of his young son from polio. Originally written in German, it was then translated into English and it is the English version that has been widely distributed. However, the original version in German reveals more accurately Einstein's intended meaning. See: https://www.thymindoman.com/einsteins-misquote-on-the-illusion-of- feeling-separate-from-the-whole/

194 Thomas Berry, *The Dream of the Earth*, Sierra Club Books, 1988.

195 Robert Bly (trans.). 1977. *The Kabir Book,* Boston: Beacon Press, p. 11.

196 Op. Cit., Swimme, *The Hidden Heart of the Cosmos.* https://en.wikiquote.org/wiki/Brian_Swimme

197 Rupert Spira, see: https://non-duality.rupertspira.com/home

198 Eckhart Tolle, *Present Moment Reminder*, November 10, 2019. https://www.eckharttolle.com/

199 Sean D. Kelly, "Waking Up to the Gift of 'Aliveness,'" *The New York Times*, Dec. 25, 2017. https://www.nytimes.com/2017/12/25/opinion/aliveness-waking-up-holidays.html

200 Howard Thurman, https://www.goodreads.com/quotes/6273-don-t-ask-what-the-world-needs-ask-what-makes-you

201 Saint Teresa of Avila, Brainy Quote. https://www.brainyquote.com/quotes/saint_teresa_of_avila_105360

202 Vandana Shiva, Earth Democracy: Justice, Sustainability, and Peace, North Atlantic Books, October, 2015. https://www.goodreads.com/work/quotes/2953610-earth-democracy-justice-sustainability-and-peace

203 Jem Bendell, "Climate despair is inviting people back to life," Posted in his blog on deep adaptation, July 12, 2019. https://jembendell.com/

204 See Dziuban's website: www.PeterDziuban.com

205 Peter Dziuban, "The Meaning of Life Is Alive," *Excellence Reporter*, November 26, 2017. https://excellencereporter.com/2017/11/26/peter-dziuban-the-meaning-of-life-is-alive/

206 Henri Nouwen, *The Way of the Heart: Connecting with God through Prayer, Wisdom, and Silence*, Harper Collins, 1981.

207 See the chapter, "Is Humanity Growing Up?" which explores these themes at length: Duane Elgin, Chapter One in the book *Promise Ahead*, Harper/Quill, 2000. Also:

Tim Elmore, "The Marks of Maturity," originally published in *Psychology Today*, November 14, 2012. http://www.markmerrill.com/the-7-marks-of-maturity/

208 Jessica Xiao, "Why Climate Change Is an Issue of Social Justice for Vulnerable People," *Everyday Feminism*, October 9, 2016. https://everydayfeminism.com/2016/10/climate-change-for-vulnerable-ppl/

209 Maya Angelou, *Letter to My Daughter*, Random House, 2008.

210 Toni Morrison, "2004 Wellesley College commencement address," published in *Take This Advice: The Best Graduation Speeches Ever Given*, Simon & Schuster, 2005.

211 See, for example: Joseph V. Montville, "Psychoanalytic Enlightenment and the Greening of Diplomacy," *Journal of the American Psychoanalytic Association*, Vol. 37, No. 2, 1989. Also:

Roger Walsh, *Staying Alive: The Psychology of Human Survival*, Boulder Colorado: New Science Library, 1984.

212 Martin Luther King, Jr., https://www.brainyquote.com/quotes/martin_luther_king_jr_101309

213 Alan Paton, https://www.azquotes.com/author/11383-Alan_Paton

214 See, for example: Dana Meadows, et. al., *Beyond the Limits,* Chelsea Green Publishing Co., 1992.

215 Tatiana Schlossberg [An interview with Narasimha Rao, professor at Yale], "Taking a Different Approach to Fighting Climate Change," *The New York Times*, November 7, 2019. https://www.nytimes.com/2019/11/07/climate/narasimha-rao-climate-change.html Also:

Environmental and Climate Justice Program, NAACP, https://www.naacp.org/environmental-climate-justice-about/

"Climate justice," Wikipedia, "A fundamental proposition of climate justice is that those who are least responsible for climate change suffer its gravest consequences." https://en.wikipedia.org/wiki/Climate_justice

216 Pedro Conceição, et al, "Human Development Report: Beyond income, beyond averages, beyond today: Inequalities in human development in the 21st century," *UNDP*, 2019 http://hdr.undp.org/sites/default/files/hdr2019.pdf

217 "Forced from Home: Climate-fueled displacement," *Oxfam Media Briefing,* December 2, 2019. https://oxfamilibrary.openrepository.com/bitstream/handle/10546/620914/mb-climate-displacement-cop25-021219-en.pdf "Countries that contribute the least to greenhouse gas emissions will likely continue to experience the greatest consequences due to climate change. The greatest impact of climate change will occur in poor countries." Also:

Barry Levy, et. al., "Climate Change and Collective Violence," *Annual Review of Public Health*, January 11, 2017. doi: 10.1146/annurev-publhealth-031816-044232

"Environmental & Climate Justice," NAACP, 2019. https://www.naacp.org/issues/environ mental-justice/

218 Desmond Tutu quoted in Terry Tempest Williams, *Two Words*, Orion, Great Barrington, MA, Winter 1999.

219 These examples were drawn, in part, from: Emily Mitchell, "The Decade of Atonement," *Index on Censorship*, May/June 1998, London (and reprinted in the *Utne Reader*, March-April 1999).

220 John Bond, "Aussie Apology," *Yes! A Journal of Positive Futures*, Bainbridge Island: WA, Fall 1998.

221 Ibid.

222 Eric Yamamoto, *Interracial Justice: Conflict and Reconciliation in Post-Civil Rights America*, New York University Press, 1999.

223 Ted MacDonald & Lisa Hymas, "How broadcast TV networks covered climate change in 2018," Media Matters, March 11, 2019. https://www.mediamatters.org/donald-trump/how-broadcast-tv-networks-covered-climate-change-2018

224 Tibi Pulu, "Mainstream US broadcast networks decrease climate change coverage time by 66% to only 50 minutes. For a whole year," *Climate News*, March 27, 2017. https://www.zmescience.com/ecology/climate/climate-change-coverage-432432/

225 For information about the "Extinction Rebellion," see: https://rebellion.earth/

226 "World Internet Users and 2019 Population Stats," Miniwatts Marketing Group, October 4, 2019. https://www.internetworldstats.com/stats.htm Some of the key takeaways from their Global Digital Report 2019 include: The number of internet users worldwide in 2019 is 4.388 billion, up 9.1% year-on-year. The number of social media users worldwide in 2019 is 3.484 billion, up 9% year-on-year. The number of mobile phone users in 2019 is 5.112 billion, up 2% year-on-year basis. See: https://hootsuite.com/pages/digital-in-2019 Also: https://wearesocial.com/blog/2019/01/digital-2019-global-internet-use-accelerates

227 Regarding access to television: "For the first time, more than half of the world's population with TV sets are now within reach of a digital TV signal. The figure stands at approximately 55% as of 2012, compared to just 30% in 2008, according to *ITU's* annual "Measuring the Information Society, 2013." Also:

Tom Butts, "The State of Television, Worldwide," *TV Technology*, December 6, 2013. https://www.tvtechnology.com/miscellaneous/the-state-of-television-worldwide With regard to TV households: Global digital penetration climbed from 40.4% of TV households at end-2010 to 74.6% by end-2015, according to the latest edition of the *Digital TV World Databook*. About 584 million digital TV homes were added in 138 countries between 2010 and 2015. This doubled the digital TV household total to 1,170 million.

According to *Digital TV Research*, "Three Quarters of global TV households are now digital," May 12, 2016 https://www.digitaltvnews.net/?p=27448

The number of TV households worldwide is projected to increase from 1.74 billion in 2023, up from 1.63 billion in 2017.

"Number of TV households worldwide from 2010 to 2018," Statista, December 4, 2019. https://www.statista.com/statistics/268695/number-of-tv-households-worldwide/

As further context: In July 2012: The world has 7 billion people, and they live in 1.9bn households, which on average have 3.68 people in them each. Of those 1.9bn households, only 1.4bn households have a TV, let alone the internet. https://www.theguardian.com/media/blog/2012/jul/27/4-billion-olympic-opening-ceremony

228 A.W. Geiger, "Key Findings about the online news landscape in America," *Pew Research Center*, September 11, 2019. https://www.pewresearch.org/fact-tank/2019/09/11/key-findings-about-the-online-news-landscape-in-america/_Perspective on the U.S. experience is found in a Pew Research study found that in 2019, 49% of Americans get their news often from television, 33% from online websites, 26% from radio, 20% from social media, and 16% from print newspapers.

229 Pocket Neighborhoods; see: http://pocket-neighborhoods.net/ Also: https://neighborhoodhomes.org/

230 Ecovillage; see: https://en.wikipedia.org/wiki/Ecovillage Also:

"Global Ecovillage Network": https://ecovillage.org/

https://www.ic.org/directory/ecovillages/

In the United States: https://www.transitionus.org/transition-towns

EcoDistricts. https://ecodistricts.org/ "Within every neighborhood (or district) lies the opportunity to design truly innovative, scalable solutions to some of the biggest challenges facing city makers today: income, education and health disparities; blight and ecological degradation; the growing threat of climate change; and rapid urban growth. EcoDistricts is advancing a new model of urban development to empower just, sustainable, and resilient neighborhoods. [EcoDistricts are a] . . . collaborative, holistic, neighborhood-scale approach to community design to achieve rigorous, meaningful performance outcomes that matter to people and planet."

231 Transition Towns refer to grassroot community projects that aim to increase self-sufficiency to reduce the potential effects of peak oil, climate destruction, and economic instability. See: https://en.wikipedia.org/wiki/Transition_town Also:

https://transitionnetwork.org/ Here is a list of transition "hubs" around the world: https://transitionnetwork.org/transition-near-me/hubs/

232 See the work of "Ecocity Builders." https://ecocitybuilders.org/what-is-an-ecocity/ They offer this description of ecocities: "Like living organisms, cities and their inhabitants exhibit and require systems for movement (transport), respiration (processes to obtain energy), sensitivity (responding to its environment), growth (evolving/changing over time), reproduction (including education and training, construction, planning and development, etc.), excretion (outputs and wastes), and nutrition (need for air, water, soil, food for inhabitants, materials, etc. Also:

https://en.wikipedia.org/wiki/Sustainable_city See how sustainable cities fit within the United Nations "sustainable development goals." https://www.un.org/sustainabledevelopment/cities/

For European sustainable cities, see: http://www.sustainablecities.eu/

233 Eco-civilizations: See: https://en.wikipedia.org/wiki/Ecological_civilization_ Pressure is building to take radical action to de-carbonize the economy as the window for mitigation is closing. A substantial emission reduction is needed before 2030 if global warming is to be kept below 2°C. Several countries have started to change policy and are about to transition into eco-civilizations, changes supported by benefits beyond mitigating climate change (e.g. health benefits). China is a world leader. Also:

"Eco-civilization: China's blueprint for a new era." https://www.theclimate group.org/sites/default/files/archive/files/China-Ecocivilisation.pdf

234 For reference, see endnotes 29 and 30 that summarizes key data sources that describe the potential for abrupt, rapid shifts in climate configurations.

235 "Life Beyond Growth: The history and possible future of alternatives to GDP-measured Growth-as-Usual," Alan AtKisson, Stockholm, Sweden, January 31, 2012. http://www.oecd.org/site/worldforumindia/ATKISSON.pdf Even these estimates may underestimate the cost of climate change. Also:

Naomi Oreskes and Nicholas Stern, "Climate Change Will Cost Us Even More Than We Think," *The New York Times*, Oct. 23, 2019. https://www.nytimes. com/2019/10/23/opinion/climate-change-costs.html

236 See, for example, the Swedish word "lagom" which means "just the right amount," "in balance," "perfect-simple." https://en.wikipedia.org/ wiki/Lagom

237 Arnold Toynbee, *A Study of History*, (Abridgement of Vol's I-VI, by D.C. Somervell), New York: Oxford University Press, 1947, p. 198.

238 Robert McNamara, former President of the World Bank, defined "absolute poverty" as: "a condition of life so characterized by malnutrition, illiteracy, disease, high infant mortality and low life expectancy as to be beneath any reasonable definition of human decency."

239 For various definitions, see, Elgin, *Voluntary Simplicity*, Op. Cit., (first edition, 1981), p. 29.

240 Buckminster Fuller describes this process as "ephemeralization." However, unlike Toynbee, Fuller's emphasis was on designing material systems to do more with less rather than the co-evolution of matter and consciousness. See, for example, his book, *Critical Path*, New York: St. Martin's Press, 1981.

241 Matthew Fox, *Creation Spirituality,* San Francisco: HarperSan Francisco, 1991.

242 Francis J. Flynn, "Where Americans Find Meaning in Life," *Pew Research Center*, November 20, 2018, https://www.pewforum.org/2018/11/20/ where-americans-find-meaning-in-life/ Also:

"Research: Can Money Buy Happiness?," *Stanford Business*, September 25, 2013. https://www.gsb.stanford.edu/insights/research-can-money-buy-happiness

"Can Money Buy You Happiness?" Andrew Blackman, *Wall Street Journal*, November 10, 2014. Research shows how life experiences give us more lasting pleasure than material things is found here: https://www.wsj.com/articles/can-money-buy-happiness-heres-what-science-has-to-say-1415569538

Sean D. Kelly, "Waking Up to the Gift of 'Aliveness,'" *New York Times*, December 25, 2017. https://www.nytimes.com/2017/12/25/opinion/aliveness-waking-up-holidays.html

243 Op. Cit., "Can Money Buy You Happiness?" Andrew Blackman.

244 Foa Inglehart, et al., "Development, Freedom, and Rising Happiness: A Global Perspective (1981–2007)," *Association for Psychological Science*, Vol. 3, No., 4, 2008. Also:

Ronald Inglehart, "Changing Values among Western Publics from 1970 to 2006," *West European Politics*, January–March 2008.

245 Ralph Waldo Emerson. See: https://philosiblog.com/2013/06/10/the-only-true-gift-is-a-portion-of-yourself/

246 With gratitude to the Buddhist monk, Thich Nat Hanh for offering this description.

247 Duane Elgin, "Revitalizing Democracy Through Electronic Town Meetings," *SPECTRUM: Journal of State Governments*, Spring 1993. https://duaneelgin.com/wp-content/uploads/1993/12/ETMs-Spectrum-Journal.pdf Also:

Duane Elgin, "The Power of a 'Community Voice' Movement," *HuffPost*, November 18, 2011. https://www.huffpost.com/entry/future-occupy-movement_b_1100549

248 Duane Elgin and Peter Russell on "Pete and Duane's Window," *Take Back the Airwaves part 2*, January 19, 2011. https://www.youtube.com/watch?v=a53hL5Z1WHE&feature=youtu.be

249 In the US, the rights of the public are profound when it comes to use of the airwaves for both radio and television. These rights are established in the Bill of Rights and Constitutional law. The First Amendment in the Bill of Rights states that: *"Congress shall make no law . . . abridging the freedom of speech . . . or the right of people to peaceably assemble, and to petition the Government for a redress of grievances."* In other words, no law will be passed that limits the right of citizens to assemble peacefully, speak freely, and petition the government for redress of grievances. This is exactly what is involved in an Electronic Town Meeting in the modern era: Citizens assemble peacefully. They speak freely. And, if there is a working consensus, they can directly petition the government asking for redress—or for setting matters right or establishing appropriate remedies.

Turning from Constitutional law to media law in the US, we find that **the public at the "local level" is the owner of the airwaves used by television broadcasters. The local level is the scope of the media footprint of broadcasters, which is generally the metropolitan scale.** Even if broadcasters use the internet to deliver much of their own programming, if they are also using the airwaves, they still have a strict legal obligation "to serve the public interest, convenience, and necessity."

Nearly a century ago, the Radio Act of 1927 established the basic rules for operation using the public's airwaves, stating that: *". . . broadcast stations are not given these great privileges by the United States Government for the primary benefit of advertisers. Such benefit as is derived by advertisers must be incidental and entirely secondary to the interest of the public."* The Commission further stated that: *"**The emphasis must be first and foremost on the interest, convenience and necessity of the listening public, and not on the interest, convenience, or necessity of the individual broadcaster or advertiser.**"*

A Federal Appeals Court clarified the role of citizens in 1966, saying: *"Under our system, the interests of the public are dominant. . . . Hence, individual citizens and the communities they compose **owe a duty to themselves and their peers to take an active interest in the scope and quality of television service the stations and networks provide. . . . Nor need the public feel that in taking a hand in broadcasting they are unduly interfering in the private business affairs of others.** On the contrary, their interest in television programming is direct and their responsibilities important. **They are the owners of the channels of television—indeed, of all broadcasting.**"* [emphasis added]

A 1969 Supreme Court decision further clarified the responsibilities of broadcasters. The court ruled that: *"**It is the right of the viewers and listeners, not the right of the broadcasters, which is paramount.**"* [emphasis added] The Communications Act of 1934 was updated by the US Congress in 1996. The resulting *Telecommunications Act* is over 300 pages long and, throughout, the principle is affirmed that **the airwaves should be used *"to serve the public interest, convenience, and necessity." Television broadcasters have no property rights with regard to using the airwaves; they only have the privilege of using the airwaves as long as they are serving the public interest, convenience, and necessity.*** [emphasis added]

Importantly, we have moved beyond a time of serving the "public interest." Given that local communities are threatened by climate change, as well as the viability of the entire planet, **we have moved to a much higher standard for broadcasters; namely that they serve the "public interest" and the "public necessity."** [emphasis added]

In practical terms, this means that if the local public (the metropolitan scale of the broadcaster's media footprint) asks for a reasonable amount of airtime to be devoted to the climate challenge (which threatens a local community, as well as the entire Earth), then the public should expect the support of the government (the Federal Communications Commission) to uphold such requests that clearly serve the public interest and necessity.

Similarly, **if the public requests airtime for Electronic Town Meetings to consider threats such as climate change, these requests for using the airwaves (which we citizens own) are entirely legitimate and are grounded in both Constitutional law and in nearly a century of Federal law.**

250 "Number of Olympic Games TV viewers worldwide from 2002 to 2016," *Statista*, 2020. https://www.statista.com/statistics/287966/olympic-games-tv-viewership-worldwide/